A HANDBOOK OF SYSTEMS ANALYSIS

J. E. BINGHAM, A.C.I.S.
Senior Consultant, Diebold Europe S.A.(Frankfurt)

and

G. W. P. DAVIES, M.A. (Cantab.)
Formerly Director of Technical Services, Diebold Europe S.A. (London)

MACMILLAN

© J. E. Bingham and G. W. P. Davies 1972

All rights reserved. No part of this publication may be reproduced or transmitted, in any form or by any means, without permission.

First published 1972 by
THE MACMILLAN PRESS LTD
London and Basingstoke
Associated companies in New York Dublin
Melbourne Johannesburg and Madras

Reprinted 1974

SBN 333 13886 4

Distributed in North America by Halsted Press,
A Division of John Wiley & Sons, Inc.,
New York and Toronto

ISBN 0 470-07319-5

Library of Congress Catalog Card Number 72-5935

Printed in Great Britain by
LOWE AND BRYDONE (PRINTERS) LTD,
Haverhill, Suffolk

A HANDBOOK OF SYSTEMS ANALYSIS

FOREWORD

By 1975 we can expect half a million persons in Europe and the United States to be employed in the type of work broadly labelled as 'systems analysis'. Yet this new profession did not exist as a separate discipline until the late nineteen fifties. The major impetus for the remarkable growth in demand for this kind of knowledge came from the rapidly increasing use of computer systems in business applications. With this came a widespread recognition that business procedures could almost universally be substantially improved by taking a hard look at their basic functions. A good portion of such improvements is due to employment of the discipline rather than the instrument. To apply the approach successfully, however, required a combination of established analytical techniques and the new technology of data processing; it was to fulfil this requirement that the new discipline emerged.

Surprisingly, the rapid growth of systems analysis has been accomplished in spite of a severe shortage of practical literature on the subject. Considerable information exists about scientific applications of the computer and indeed about the theory of systems analysis itself. What has been lacking is a book which gives practical guidelines for the systems analyst or trainee in the normal business environment, which after all accounts for around 90% of all computer installations.

It was therefore with pleasure that I accepted the invitation to write the foreword to this book, because it eminently fills that gap. Our whole progress in Western civilisation is in danger of clogging up through paperwork, inefficient procedures and escalating costs, because of our amazing ability to mis-use the computer. Only if we attack with vigour and urgency the problem of developing a sound approach to systems analysis will we overcome this deficiency. I particularly welcome this book therefore as it aims exactly at fulfilling that need, namely a practical handbook based on solid experience.

The pleasure is all the greater for me because of the close personal and professional relationship I have enjoyed with the authors over a number of years. Thus I can vouch for the thoroughness of their work and for the sound experience on which the book is based and which is evident in the text. Indeed, it may well represent a milestone in the development of this important but hitherto ill-defined discipline we call 'systems analysis'.

September, 1971.   Henry F. Sherwood Vice President
Diebold Europe S.A.   Frankfurt/Main

CONTENTS

Acknowledgements	xi
Introduction	1

PART I The Six Steps of Systems Analysis

1	The Structure of Systems Analysis	5
2	Step 1 : System Project Selection	6
	2.1 Sources for Ideas for Systems Projects	6
	2.2 Criteria for Selecting Systems Projects	8
3	Step 2 : The Feasibility Study	11
	3.1 The Cost-benefit Analysis	12
4	Step 3 : Definition Phase	22
5	Step 4 : Design Phase	26
6	Step 5 : Implementation Phase	33
7	Step 6 : Evaluation Phase	35
8	Review of the Six Steps of Systems Analysis	37

PART II *Techniques*

9	Fact Gathering	43
	9.1 Interviewing	44
	9.2 Other Fact-finding Techniques	46
10	Flowcharting	52
	10.1 The Technique	52
	10.2 Use of the Technique	53
	10.3 Drawing a Flowchart	53
	10.4 Automated Flowcharting	58
11	Decision Tables	61
12	Simulation	68
	12.1 Desk Simulation	71
	12.2 Benchmarks	72
	12.3 Simulation Entirely by Software	74

PART III *General Systems Considerations*

13	Data Capture and Input/Output	81
	13.1 General Design Considerations	81
	13.2 Input Methods	85
	13.3 Output	93

14	File Organisation and Record Layout	96
	14.1 File Organisation Structures	99
	14.2 Facts Governing the Choice of File Organisation	104
	14.3 File Organisation Terminology	115
15	Data Security	120
	15.1 Physical Security	120
	15.2 Systems Controls	121
	15.3 Audit Controls	127
16	Standards and Documentation	128
	16.1 Performance Standards	131
	16.2 Sources of Standards	132
	16.3 Introduction and Enforcement of Standards	133
	16.4 Standards Must Enjoy Management Support	135
17	Data Communications	152
	17.1 Data Collection and Transmitting Equipment	153
	17.2 Communications Link	156
	17.3 The Receiving Unit	157
	17.4 Data Communications and Systems Design	158
18	Systems Maintenance	161
	18.1 User Involvement	162
	18.2 Modularity of Systems	162
	18.3 Modification Procedure	163
	18.4 Organisation	164
19	User Involvement	166
	19.1 Top Management	167
	19.2 Line Management	167
	19.3 Non-management Personnel	168

PART IV Project Control

20	Problems of Implementation	171
	20.1 Personnel Training	173
	20.2 File Creation and Conversion	174
	20.3 Programming	175
21	Planning for Implementation	176
22	Management Control	182

Selected Bibliography   186

ACKNOWLEDGEMENTS

Publication of a book of this nature is in many respects a team effort and we would like to acknowledge all the assistance we have received in creating it. We would like to thank the Editor of *Business Systems* for permission to reproduce the substance of Chapter 21, which has already appeared as an article in that journal. We would also like to record our gratitude to our colleagues in Diebold Europe for the many fruitful discussions we have had on subjects included in this book.

Our greatest debt is, however, to our wives who typed the entire manuscript in its several drafts and whose patience in accepting the long hours necessary for the writing of this work enabled it to reach completion.

J.E.B.
G.W.P.D.

INTRODUCTION

This book is about the analysis of business systems—not about computers. In many cases, of course, this highly valuable tool will be the means selected to implement the solution to a business problem. Indeed, the growth of systems analysis is closely related to the growth in the use of computers. Nevertheless, the *principles* of systems analysis are independent of the mechanisms used to apply them. How, then, does systems analysis differ from the longer established organisation and methods (O & M) approach? Or for that matter from any other problem-solving discipline? Cannot the approach of a mathematician be used? Or a psychologist? Or an economist? The answer to these questions is that systems analysis is in fact an extension of traditional problem-solving disciplines which have already been used in the business area. However, the very fact that systems analysis uses the approaches of the mathematician, the psychologist, the economist and the O & M man gives rise to the need for the separate discipline of systems analysis. There is clearly a need to combine those elements of the different approaches which are best suited to the solution of problems in the business environment, oriented, though not exclusively, to the use of computers.

The objective of this book is to establish a methodology for this discipline; a pragmatic approach to the identification of problem areas, their analysis and the design of practical, economic solutions. In addition to outlining this methodology this book identifies and provides working guidelines for the major techniques involved together with practical advice on their application. The vital topics of planning and the actual implementation of a new system are also discussed to provide a comprehensive treatment of the subject.

Obviously, all these topics cannot be treated exhaustively in a volume of this size, especially as systems analysis by its very nature frequently overlaps with a number of other disciplines. This text, therefore, provides a framework which further study and experience can expand into a well developed knowledge of a wide-reaching subject. In addition to its use as a primary text for newcomers to the field, it will also prove valuable as a guidebook to practising analysts whose duties have not hitherto required a knowledge of specific aspects of the discipline. This book will provide such readers with an outline of all the major areas and techniques and perhaps more importantly place them in the perspective of the discipline as a whole.

For the general reader, the book—although not written for that purpose—will provide a valuable insight into the work necessary before the power of a computer can be harnessed within a normal commercial or governmental environment.

The result of reading this book will, we hope, be the removal of some of the mystique which currently surrounds the subject and the substitution of a solid, well-founded methodology which has already been proven in the field. The systems analyst is indeed fortunate that all his work is subjected to a very simple absolute test: *does it work?* It is our hope that a careful study of this work will increase the percentage of satisfactory replies to this question!

PART 1

The Six Steps of Systems Analysis

1
THE STRUCTURE OF SYSTEMS ANALYSIS

If you ask the average systems analyst to describe and define the work he does, you will probably receive a very vague answer. This is not because his work is particularly complex or difficult to describe (it is NOT, as this book attempts to show), but simply because most systems analysts themselves do not have a very clear idea of the structure of their task. They are usually heavily overworked and manage to do their daily work because in most companies there is always a backlog of systems work which is clearly defined due to its urgency. So they follow a series of short-term goals rather than a *methodology* of systems analysis.

This book purports to show that there *is* a methodology of systems analysis, that systems analysis itself can be analysed into clearly recognizable steps. Each of these steps can be defined in terms of activities, so that one obtains a *structure* for the task. There is nothing mysterious about this. Indeed, the authors feel that the book will have been a success if it removes some of the mystique which surrounds systems analysis.

It is not claimed that every aspect of systems analysis can be defined in terms of the six broad steps described in this Part, because there are a number of other important aspects, like standards and documentation, which relate to all the steps. These are treated later in the book. Nevertheless, if a novice systems analyst follows the six steps described hereafter, on completion of a given task he will have done a thorough job of systems analysis.

The six main steps are:

1 System Project Selection
2 Feasibility Study
3 Definition Phase
4 Design Phase
5 Implementation Phase
6 Evaluation Phase

We will first describe each of these separately and then relate them to each other, to show how together they form a carefully ordered approach.

2
STEP 1: SYSTEM PROJECT SELECTION

Every systems project must at some stage have been the object of a selection process. Usually the systems analyst does not take significant part in this process and in many cases he first becomes involved when the decision has already been made. Nevertheless, the systems analysis function begins in the first place by identifying the area for study, even if this phase is often undertaken by general management rather than by systems analysts.

Before selecting a project, it is of course necessary to establish what projects there are to choose from. So let us first consider the possible sources for ideas for systems projects.

2.1 Sources for Ideas for Systems Projects

2.1.1 *Requests from Other Departments*

A very common source for ideas for projects is requests or suggestions from departments within the company. This is very useful, as these are the people for whom one wants to provide a service. But one should beware that what is convenient for one department may not be in the best interest of the company as a whole. For example, perhaps the cost accounting department would like to receive special weekly statistics of costs of materials used at each cost centre throughout the factory. In itself this may be a good idea, but to implement such a system may involve creating special sets of codes to identify the materials in the way that the cost accounting department wants. This may conflict with and partially duplicate the existing codes which are in standard use by the production and purchasing departments. To introduce an additional set of material codes would certainly cause confusion and very probably extra work for the personnel who have to collect the data for processing. In such a case, therefore, the systems department would do its utmost to try and find a way to satisfy the cost accounting department's requirements with the existing set of codes.

2.1.2 *Opportunities Created by New Hardware, Software or Techniques*

Sometimes a new technical development makes possible a new approach to a problem. Examples of this are optical character recognition, cheaper

STEP 1: SYSTEM PROJECT SELECTION   7

data communications, etc. Genuine advances may be made in this way, but the point to watch is that one should not take a solution and then search for a problem. This may lead to subtle, clever systems which do not, however, satisfy the users of the system (or solve their problems), who are left to work with the system long after the systems department has moved on to even more interesting techniques.

2.1.3   *Ideas from Previous Systems Studies*

This is a major source for systems projects. At the end of any project one is usually left with a host of ideas for better ways to have done the job just completed. It is important to document these ideas at the time to make them available when possible projects are being reviewed. For example, perhaps the systems analyst has been working on a system designed to help production personnel schedule jobs to run on the different machines available. Although the scope of the system may have been limited to this broad objective, the analyst may well have recognised that the information generated by this system would also provide the machine maintenance department with vital information for their work. So it would be sensible for the systems analyst to note his ideas at the time so that when the resources become available a deeper investigation can be made.

2.1.4   *Possibilities for Linking or Integrating with Other Systems*

This source is frequently related to point 2.1.3 above. During a systems project it can often be seen that a linkage exists with another system. Sometimes these linkages are so many and so complex that the systems are best considered as one *integrated* system. In the example described in the previous paragraph, the machine scheduling system can be seen as just one part of a much wider system ranging from the initiation of each customer order through production planning to final despatch of the goods. In such a case, it may be decided to link these different elements together to form an order entry, processing and despatch system. This viewpoint may emerge from development work in isolated areas, such as the machine scheduling system.

2.1.5   *Outside Sources*

Many systems analysts base the systems they design exclusively on the situation within their own organisation. This is a mistake. The requirements of one company are rarely unique and an exchange of ideas with other concerns either directly or through trade literature, seminars and professional organisations is always beneficial to all concerned and this is (or should be) a fruitful source of ideas for systems projects.

Through one, some or all of the sources listed above, a list of apparently possible projects is obtained. The task then becomes one of selecting from this list those projects which are in the best interests of the company.

2.2 Criteria for Selecting Systems Projects

The factors listed here are all dependent on the *judgement* of one or more experienced personnel in the systems department. It is to be hoped that their judgement will be confirmed where possible (for projects commenced) by the subsequent feasibility study, but it should be recognized that the selection of systems projects consists of only a very preliminary investigation into the following factors:

2.2.1 *Potential Return on Investment*

This is the most important criterion for selecting a project. A very broad estimate of the potential benefits and costs to the company gives a strong guide to the desirability of implementing a system. During the Feasibility Study (Step 2), this aspect is developed more fully and the reader is referred to that section for the treatment of cost/benefit analysis.

2.2.2 *Management Desire*

Occasionally, senior management become very enthusiastic about a project, even when it does not represent a particularly good financial investment. It is often then prudent to implement the desired system to maintain and gain management interest in systems work. As discussed in Part IV of this book, management involvement is absolutely essential for the success of systems work and anything within reason that can be done to develop this involvement is worth while. An example of such a project is known to the authors, where the chairman of the board himself requested a video display terminal on his desk to display various financial and personnel statistics of a very simple nature, which could more easily and cheaply have been provided manually. Although the need for such an expensive data retrieval system was very doubtful and certainly not justifiable from a cost-effectiveness point of view, the systems department wisely implemented such a system and have earned the ever-increasing support of the chairman for their activities ever since!

2.2.3 Technical Feasibility

Although a project is sometimes desirable from many points of view, it is sometimes clearly technically infeasible. An example of this is an optical character reading (OCR) system which would depend on the machine recognition of handwritten characters produced by personnel in different countries, as may happen in the case of a multi-national company. On a local or more limited scale such a system may be feasible, but to hope to overcome the problem of identifying by machine the different scripts of different nationalities would be too optimistic as any OCR specialist would admit. In such cases, a quick assessment of the technical feasibility of the project will reveal its unsuitability.

2.2.4 Capability of the Systems Department to do the Project

It is a sad fact that many apparently desirable projects are allocated to the systems department without any consideration being given to their capability to do the job. We refer not only to their frequent inability to fit the project into an already full work schedule, but also to the lack of technical knowhow which some projects demand. Real-time projects, which always involve complex use of the computer, are particularly prone to this type of failure. In such cases, it is usually necessary to hire the assistance of consultants or to recruit expertise, both of which increase the cost of the project, which should be taken into account when the cost benefit analysis is calculated. The capability of the systems department to do the job should always therefore be considered amongst the principle criteria for selecting a project.

2.2.5 Requirement for Integration with Other Systems

It is frequently necessary to undertake a project simply because *other* projects cannot start or be completed unless the project in question is accomplished. For example, a quality control information system may be highly advantageous to the company, but it depends so heavily for its input on production control information that this latter system has to be developed first, even though in itself it may not be profitable.

2.2.6 Critical Company Need

Sometimes a company finds itself in the position of urgently needing an information system in a particular area, even though the financial benefits and other justifications are unclear. Obviously if the need is felt to exist, the benefit is there somewhere and is probably ultimately measurable in financial terms. But the major justification for the project at the time is expressed simply in terms of urgent need. An example of this is the opening up of a new production plant, where the actual benefits of a computerised information system are not known. Nevertheless, it may be clear that to allow the new plant to operate tolerably efficiently depends upon the provision of an information system. A similar situation may arise when all the company's competitors implement a particular system thus forcing, for reasons of continued competitiveness, implementation of a similar system, too.

At the completion of this phase of project selection the projects to be embarked upon should have been defined, each with a clear statement of its objective. This is essential for the meaningful completion of the following phase : The Feasibility Study.

3
STEP 2: THE FEASIBILITY STUDY

The objective for this phase is a report indicating the possible ways of accomplishing the project objective, stating the costs and benefits of each approach. Normally, a recommendation for the best way is implicit, if not explicit. Here are the main activities comprising the feasibility study :

- Identify the main characteristics of the system.
- Determine the main output requirements, including response times.
- Analyse the organisation chart, geographical distribution, etc., of the department(s) involved.
- Determine the varieties of data and estimated volumes.
- Consider possible alternative ways of meeting the requirements of user(s).
- Examine other systems meeting similar requirements.
- Prepare gross estimates of probable overall implementation and operation costs for each possible alternative.
- Prepare gross estimates of probable overall direct and indirect benefits for each possible alternative.
- Document the Feasibility Study in a report for user and systems management.
- Determine that the requirements of the system are consistent with the company objectives.

As indicated above, the main ingredient of the report documenting the Feasibility Study is the *Cost-Benefit* analysis. This is a fundamental piece of information for the management of both the systems department and the user area, and, as such, must be accurate and clearly understandable. Unfortunately, many companies do not have standard ways of handling this, each analyst using his own method of evaluation, which he himself may well vary from project to project. There is no reason why perfectly straightforward methods cannot be used in every case and indeed most systems analysts would welcome this. In the following section we therefore hope to provide a basis for such an approach by pinpointing the key areas of the cost-benefit analysis and presenting a standard method of handling them.

3.1 The Cost-Benefit Analysis

There should be three sections in any cost-benefit analysis : 3.1.1 Costs, 3.1.2 Benefits; 3.1.3 Return on Investment.

3.1.1 *Costs*

These occur in a number of different areas. It is important to consider *all* the implications of introducing a new system. A cost analysis should therefore include the following areas :

- Systems Personnel.

Include not only salary, plus overhead costs of the team members, but also any special materials, extra office space required, computer trial runs, visits to other companies, etc. Do not forget the costs of the Feasibility Study itself!

- User Personnel.

These should include the time taken up by the users to supply the information needed for the study, the time needed to implement the possible solutions (for example, setting up new files, training costs) and the time needed to maintain the system once implemented. Implementation of the chosen solution will almost certainly require the writing of new computer programs. These are normally a large part of the overall project cost. Estimates of the degree of maintenance needed later should also be included.

- Equipment Costs.

Not only the costs for running the computer should be included here, but also the costs of storing the data in machine-coded form (for example, each magnetic tape costs around £12; disks are more expensive). If new equipment is to be introduced with the new system, then clearly the costs of this must be indicated. These can best be shown as a constant value, even where purchased rather than rented, because most companies amortise capital costs over a period of time.

- Other Costs.

The introduction of new hardware implies the provision of space and special features for it (air-conditioning, electrical power, etc.). Such costs are part of the system cost. In addition, new forms and other materials may be needed.

It is very important to consider how the above costs vary over time. The *rate* of investment is always an important factor for management, especially in relation to the rate of return, discussed later in this section. The following charts show the way the costs for each element vary with

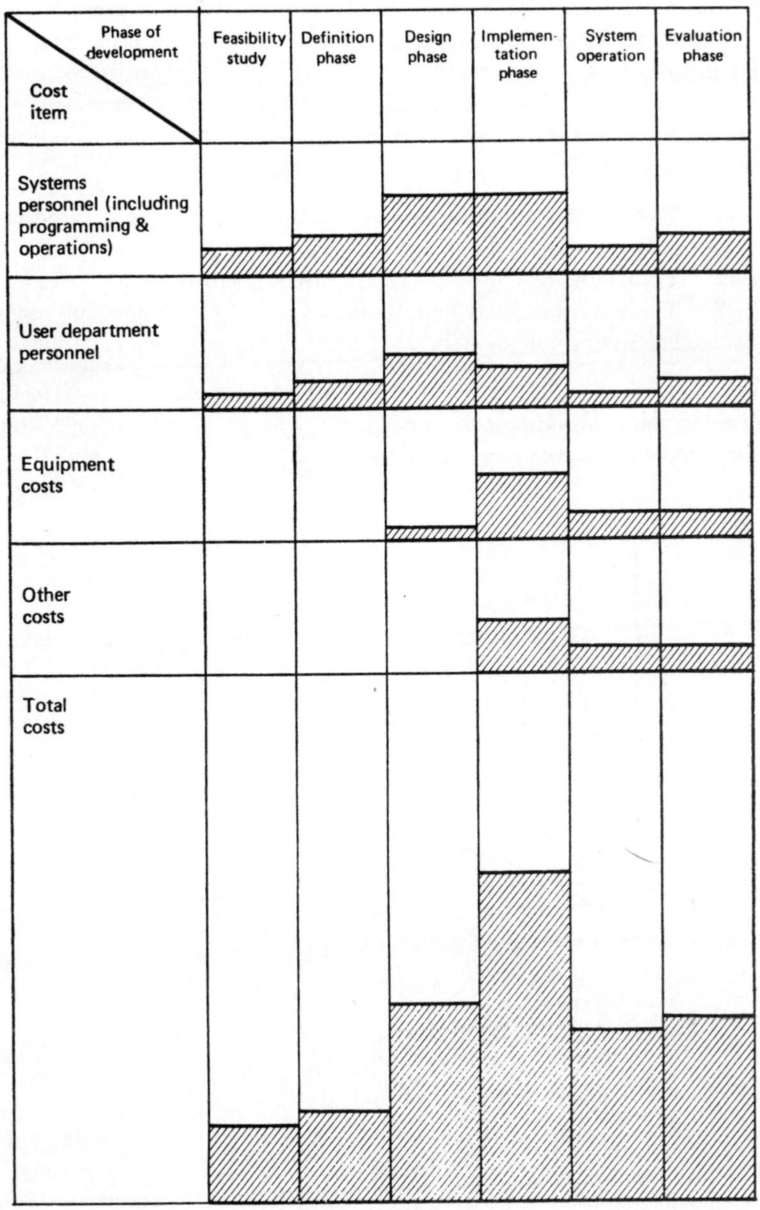

STEP 2: THE FEASIBILITY STUDY 13

Fig. 3.1. Distribution of costs of a typical system during development and operation.

time for a typical systems project and an accumulative cost/time curve has been compiled representing the sums of the other five areas.

The above figure represents a generalised case and has been simplified for this illustration. For example, the chart is based on project phases of equal time spans. This is not usually the case, but as the ratios vary from project to project it has been necessary to show a generalised case. From the curves, the following points are of significance:

(1) The variation of total costs over time are characteristic of systems projects.
(2) Total costs reach a *peak* during the implementation phase.
(3) There is a relatively high residual *operating cost*, once the system has been implemented.

We are now in a position to take the cost analysis one step further and develop the *cumulative* cost curve (against time). This is an important curve which will enable the break-even point to be determined.

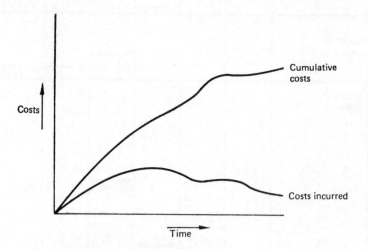

Fig. 3.2. Cumulative cost curve for the development and operation of a typical system.

A more rigorous analysis should include also the interest payable on the money needed to meet the cumulative costs. Even if the money is not in fact borrowed, there is a loss of potential interest which could be *gained* if the costs had not been incurred.

Together, the cumulative cost and cumulative benefit curves determine the all-important cumulative profit/loss curve, to be discussed later in this section.

3.1.2 *Benefits*

The benefits to be obtained from a new system fall naturally into three categories:

- Direct Savings.

These are the costs which are reduced or eliminated as a result of the new system. They are usually the main benefits aimed at in conventional systems analysis. They are rarely great, however, in comparison to the total incurred costs of the new system. Typical examples of a direct saving are:

(1) Reduction in clerical personnel.
(2) Elimination of some specific costs, e.g. postage, stationery, office machinery.
(3) Reduction in costs due to improved procedures, e.g. data capture at source may eliminate some checking of data.

Such savings are easily calculated or measurable.

- Measurable Benefits.

These are definite increases in the money accruing to the company, achieved by some feature of the new system. An example of this is the increased working capital obtained through reduction in stock levels.

Many systems analysts fail to place monetary figures on benefits, which seem intangible, but in fact can be related directly to cash gains. A good example of this is an invoicing procedure which is speeded up by a new computer-based system (a common application). Every day gained in reducing the time taken to send out invoices brings the company extra cash, which can be regarded in either of the following ways:

Either as a one-time cash gain equal to the daily billing rate times the number of days gained, or as the interest accruing from a sum equal to that previously specified.

The point here is that by sending out the invoices at an earlier time the money is also obtained more quickly (assuming that the paying habits of customers remain the same). The introduction of a new system to send out the invoices earlier therefore results in what is effectively a cash gain to the company.

Many other applications which are usually justified merely on the basis of clerical and other savings can be shown to have measurable benefit of the kind indicated above.

- Intangible Benefits.

In almost every application there are clearly desirable effects, which are difficult to evaluate in monetary terms. Very often the benefit seems best described as 'better information'. It is nevertheless sometimes possible to analyse what is meant in any one case by 'better information' and to derive a monetary value. It is always worth the effort, because benefits in this area are usually great. This fact is usually known *intuitively* by the manager who will receive the benefit and he can sometimes clearly state that his decisions will be improved by the new system. This already provides a first opening for analysis of the benefit—how can we measure better decisions?

The approach to follow is to take each decision in turn and to study in quantitative terms the effect of improved information on the decision. An improvement in the accuracy of information can be stated as an increase in the probability that the information is correct. So, for example, a manager who has to decide whether or not he should allow credit for a certain customer depends very much on his knowledge of the customer's previous payment record. If the facts are rather uncertain (say a probability of accuracy of 0.5, that is, it is a fifty-fifty chance whether or not the information is correct and complete), then he may decide not to award credit, because the risk is too great. On the other hand, if he knew the information to be complete and accurate with a probability of 0.9, then the manager may well award credit. The benefits to the company with an improved credit information system are potentially twofold :

- increased volume of sales
- reduction in bad debts.

The two are obviously closely related to the company's previous performance in this area. If they have had an 'easy' credit policy, then apparent sales may be reduced by an improved credit information system but the bad debt level will go down. So the real income to the company will tend to increase as bad debts are reduced and credit is awarded to customers, who may previously have been wrongly thought to be bad risks. In this way a monetary figure can be established for the *value* of improved information.

Let us quote two actual examples known to the authors of the way in which the value of better information can be determined. Firstly, one company has installed a cash management system, which gives instant information on the level of cash held by the company in its various bank deposits. Previously this information was known only on a weekly basis. It is now possible, however, for the company's financial management to study the cash situation at the end of each day and switch investments and allocate resources according to the most profitable return. In this way calculable (very large) benefits are achieved by the new system simply from better and earlier information.

STEP 2: THE FEASIBILITY STUDY

In the second example, a new system was developed to install a capital assets control system with the main objective of speeding up the year-end accounting process, which normally took over two months. With the new system this year-end process took only a few days. Although there were reductions in clerical work involved, these savings were fairly trivial. The main benefit was calculated to lie in the area of insurance. Under the old system the whole of the company's capital assets were insured at the previous year's value until the two months' processing task had been completed. The new system provided an almost immediate readjustment of insurance values. The value of this information here lay in two possibilities:

(a) that the company was paying too much in insurance for the value of its assets;
(b) that the company was under-insured so that in the event of loss or damage to assets it would recover insufficient compensation.

It was possible to estimate the probabilities and values of each of these. The value of having the information earlier was then found to be many, many times greater than the cost of providing it.

We hope to have shown that many apparently 'intangible' benefits can in fact often be quantified. However, there will always remain certain clear benefits which cannot be quantified. In these cases there are two steps which should be taken:

(i) A points weighting scheme should be applied.
(ii) Signatures should be obtained from interested parties that these benefits will accrue and that they agree with the importance of the benefits as indicated by the weighting scheme.

The points weighting scheme operates in the following way. A list of the intangible benefits is drawn up. Each of this is then allotted a relative value of points, with a maximum of 10 for each. By calculating the total of these and by comparing benefit to benefit, an evaluation can then be made between alternative systems. Furthermore, a criterion can be set, which states that a system having an intangible value of x points is worth y units of monetary value. Hence, it provides a basis for justification of systems which have many and/or significant intangible benefits, even if in strict financial terms no direct return on the money invested is generated.

Let us take an example of comparing two order entry systems (Fig. 3.3). This comparison would indicate that both alternatives A and B are very much preferable to the present system and that alternative A is preferable to alternative B. This also then acts as a basis for justification (benefits versus costs) of system A. System A is so much more beneficial than the present system (by a factor of 2) that it may well be decided that this difference is well worth the extra cost of installing it. It does at least give an indication of the relative values of the system.

It is important, however, to obtain the *signatures* of the management in the areas concerned that they consider the identified intangible benefits to be worth the incurred cost of the system. This exerts a strong influence on the strength of their convictions!

Intangible benefit \ System	System A	System B	Present system
Errors in resultant reports	2	6	1
Flexibility for adding new products	4	2	4
Ease of inserting urgent orders	5	5	6
Possibility to link with production planning system	8	1	0
Ease of answering customer enquiries	6	5	1
Tot	25	19	12

Fig. 3.3. Sample points weighting analysis of intangible benefits

3.1.3 *Return on Investment*

Once the analysis of costs and of benefits has been completed, the crucial test then arises of whether the combination of the two factors makes the system a worthwhile proposition to the company. The term 'worthwhile' has, of course, to be defined. If the criteria are to be strictly financial, then a clear basis for decision can be established. The usual way to do this is to state the required *payback period*.

The payback period is simply the span of time which elapses from the time the first part of the new system goes operational to the point when all invested and running costs have been recovered through the benefits of the new system. Hence we may require that all investment has been recovered within 18 months of system installation, which is the figure used by a major computer user known to the authors. This same company also requires that the break-even condition (that is monthly benefits equal monthly running cost) is reached within 9 months. This is to guard against systems from which benefits may be large but which do not realize them until a relatively late stage.

The last point is an important one. Benefits which do not appear until a distant point in the future are now recognized as being of less actual value than they at first appear. There are several reasons for this:

STEP 2: THE FEASIBILITY STUDY

(1) Constant inflation means that the real value of a given sum at a future date will be less than the present value.
(2) There is always some uncertainty about the future, and this uncertainty increases according to the distance in time of the future point. This is particularly important in the case of computer systems, which are subject to constant change. Amendments, due to both internal and external factors, may reduce or even wipe out the expected benefit of the system.
(3) Benefits obtained immediately can be invested and will therefore generate more benefits. Costs not covered now, on the other hand, require interest to be paid if the money is borrowed, or lost potential interest if it is taken from existing funds.

These facts have been recognized by financial analysts, who have developed a simple but effective way of giving a present value to future benefits. The method is known as the Discounted Cash Flow method and should belong to the repertoire of analytical techniques of every systems analyst.

The most straightforward way of applying the Discounted Cash Flow (DCF) technique is to apply a carefully chosen standard interest rate to the net cash flows (i.e. costs subtracted from benefits) estimated for each year of the life of the project. No distinction is made between capital and revenue items, nor is any account taken of such book-keeping transactions as depreciation. However, the effects of taxation and investment grants must be allowed for.

Thus if a net benefit of £100 is expected in one year's time and the discount rate is 20% the present value would be £80. If the same benefit was expected in 2 years' time the present value would be £64.

After all the estimated cash flows have been discounted to their present worth the gross negative cash flow is subtracted from the gross positive cash flow. Only if the resulting Net Present Value (NPV) is positive should further effort be expended on the project.

Of course, this will depend on the rate of discount chosen and this should reflect not only the current interest rates within the economy but also the degree of risk associated with investments in the area under consideration. Rates commonly used vary from 10–20%. After calculation the project with the highest NPV will normally be preferred.

Although the application of an annual discount rate automatically takes the increased uncertainty of future events into account, DCF analysis is, like any other analytical technique, only as good as the estimated data on which the calculation is based, and considerable effort should be made to ensure that this information is as accurate as possible. Furthermore, the calculation is fairly sensitive to the estimated life of the project and careful consideration should also be given to this factor.

However, DCF analysis is at least a quantitative way of recognizing that benefits which will arise at a point in the future have less real value than the nominal amount. This point is important in systems work, because it is usual for a considerable investment to have taken place before benefits start coming in at any significant rate. Let us look at the typical time versus cost/benefit curve for computer-based systems.

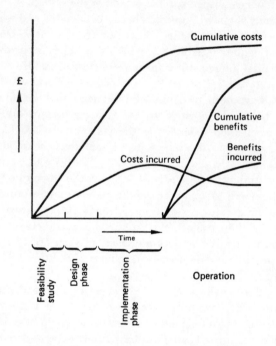

Fig. 3.4. Cost-benefit curves for typical systems.

Note that although the break-even *condition* is reached relatively soon after implementation is completed, the actual break-even point (accumulated benefits equal accumulated costs) does not occur until relatively late. As this is a generalized diagram, no actual time units have been put on it. But if we consider a modest system, which takes six months to conceive and implement, then for a cost-benefit pattern of the typical kind shown by the diagram the break-even point is not reached for two years.

This indicates that many systems do not in fact ever directly contribute to the company's profits. This is borne out by one's experience and observation of the real world. Most routine systems operating in business would not stand up favourably to the sort of cost-benefit analysis out-

STEP 2: THE FEASIBILITY STUDY

lined in this chapter. One of the reasons for this is that the profit criterion is not normally as important as one might suppose in the decision to install a system or not. Provided a system is financially justified for operating cost/benefit only, even if only marginally, most companies are prepared to install a system for other reasons, such as the desire to be 'progressive', or on the grounds that the system will provide the basis for future expansion (often doubtful). Very often only lip-service is paid to the true financial aspects of the system.

The authors are convinced that this is wrong. It may well be that the system is necessary for future expansion, but this should then be a major objective of the system and not a 'side-benefit'. If the system is to make money for the company, then this should be clearly stated and quantified so that the Feasibility Study shows clearly if the system is going to meet the company's requirements or not. The point is that money tied up in barely profitable systems is really a *loss* to the company, because it could be using that money in other areas which bring in a higher rate of return. It is important therefore to view the system from a total company point of view. A standard should be set for a minimum rate of return, so that any proposed system whose benefits fall below that level is not implemented unless the goal of the system is not financial.

4
STEP 3: DEFINITION PHASE

The objective of this phase is to obtain a *System Definition*, which will then be implemented, if accepted, in the subsequent design phase. In contrast to the Feasibility Study, detailed analysis of every aspect of the system must be carried out. The basic activities are :

- Determine objectives of current system.
- Study current system to see how far it meets its objectives.
- Analyse users' and company's requirements to develop new objectives.
- Identify constraints imposed by users' environment.
- Identify users' responsibility for data inputs and outputs to other systems.
- Examine interaction of proposed system with other systems.
- Prepare details of requirements of users—data elements, volumes, response times, etc.
- Prepare design specifications.
- Prepare plan for design and implementation phases.
- Produce a report for the user and systems management.

This phase requires working closely together with the users. An intelligent and open-minded approach is required on behalf of the analyst, who must often make judgements as to whether a user requirement is *necessary* or just *desired*. The analyst will talk to many different levels of users and will have to record and sift a large amount of data. He should aim at identifying the underlying problem and not at just recording the effects of the problem. A system that aims at alleviating the symptoms of a problem will invariably be much more complex, messy and difficult to implement than a system which gets directly at the problem itself.

The key to this phase is, therefore, to establish the *objectives* of the activities being analysed. Once these have been determined, the problems in meeting them become clear and an appropriate system to overcome the problems and meet the objectives can be defined.

Systems objectives can be defined at several levels and it is in fact desirable to do just that. Firstly, there are the overall objectives of the total information system. The essential point about these objectives is that they

STEP 3: DEFINITION PHASE

contribute towards the company objectives. Examples, therefore, of overall system objectives are:

- To maintain the company's share of the market.
- To increase the net profits of the company.
- To maintain the current efficiency of operations during expansion of the business.
- To provide an information system of a guaranteed level of sophistication at a cost which does not exceed 1% of annual sales turnover.

Although these are meaningful and important in determining systems strategy, they do not help the analyst much in his determination of objectives at departmental system level. The overall system objectives should therefore be broken down to *sub-system level*. In doing so, it is important to observe the following principles:

- The sub-system objectives must be consistent with the overall goals of the information system.
- Wherever possible the objectives should be quantifiable so that progress may be measured.
- The objectives must be achievable.
- Each sub-system objective must be capable of itself being broken down into sub-objectives.

An example of possible sub-system objectives and sub-objectives is given below:

ORDER ENTRY SYSTEM

Objectives	Sub-Objectives
Improve customer service.	• Reduce delivery time by an average of: 20% for products types A, B, C. 5% for products types D, E. • Provide immediate answers to 80% of customer enquiries. • Reduce average time to confirm acceptance of an order by 15%.
Reduce stocks of semi-finished products.	• Reduce stocks by: 40% for products F, G. 80% for products A, D. 10% for products C, H, J, K.
Provide better information for the production planning sub-system.	• Maintain in real-time a file of outstanding orders. • Maintain a daily updated file of orders shipped. • Send a daily report of order amendments to production planning.

This translation of the main objectives of a sub-system into sub-objectives expressed in more tangible terms is an essential part of the systems analyst's task. Another example, taken from the production area, is given below:

PRODUCTION PLANNING

Objectives	Sub-Objectives
To optimize the use of the plant.	● Plant availability constantly matched against outstanding orders. ● Maintenance requirements are co-ordinated with scheduling. ● Group orders for similar products together. ● Calculate optimum batch sizes.
To prevent over-production.	● To build a realistic scrap allowance ● To check for amendments to the order during production. ● To optimise order requirements against available material.
To increase the flexibility of production planning.	● Insert 80% of urgent orders into plan within 1 day. ● Be able to re-allocate jobs planned for production unit 1 to production unit 2 up to 24 hours before production is due to start.

In this way the systems analyst can identify the problem in general terms and then by careful investigation gradually break it down to lower level objectives. This step by step process, proceeding gradually into a greater level of detail, is essential, as it enables the systems analyst to become increasingly familiar with the problem area without being plunged directly into confusing detail. He will obtain initially an overview of the problems; then he can break these down into sub-problems of a manageable size.

Having defined the objectives of the system, the best approach for the systems analyst to follow is to identify all the *data elements* involved in the required output from the system. A data element is the lowest level logical unit of information. For example, a man's name is a data element, so are the price of an article, a product code, a date, etc. It is not normally meaningful to break these down further. Part III, 16 discusses the way in which data elements may be formally described.

After analysing the data elements present in the desired output from the system, it is then necessary to examine the ways in which these may be entered into the system. In doing so, certain relationships between data elements will emerge (those that are not already known) and later, during the Design Phase, this will help define the manner of data capture and of file organisation. In this way a definition of input, output and file content is developed.

Based on the identification of objectives, input, output and file content, the vital document called the 'System Definition' can be developed. This should contain:

STEP 3: DEFINITION PHASE

- Definition of objectives of the system.
- Definition of constraints on the system (for example, certain company departments may not be included).
- Narrative of system function.
- Overall information flow.
- Specification of input data and means of creation.
- Specification of output data and its distribution.
- Definition of content of files and method of updating.
- Definition of responsibilities (including data input, updating, error control, etc.).
- Definition of time constraints (for example, latest time at which information is to be received by each user).

The format of the individual documents which constitute the systems definition are further discussed in Part III, 16 on Standards and Documentation. Note that the System Definition does not indicate the way in which the system should be implemented or the amount of resources needed. It merely specifies the content of the operational system and the way in which it should be maintained. Note, too, that at this stage only the *content* of files is defined; the way in which the file is to be organized is left to the Design Phase. Also, the definition of responsibilities does not describe the *procedure* to be followed, merely where the responsibilities lie for input, updating, checking, etc. The way in which these functions will actually operate emerges in the Design Phase.

Once the System Definition has been produced the systems analyst is in a better position to estimate the time and resources that will be needed to design and implement the system. Although the number of activities identified will not be large at this stage, it is advisable to use a formal planning technique, such as a network or job progress chart, discussed in Part IV, 21, as the project will grow and grow from this point and it is highly desirable that good control is maintained from the beginning. Furthermore, one of the purposes in developing a plan at this stage is to provide the eventual user with an idea of his commitment to it and the time at which he can start to expect results.

The plan for the design and implementation phases is combined with the System Definition into a report for systems and user management. It may well be necessary to make amendments to the definition after they have reviewed it. Time used at this stage in agreeing on a definition is well spent, as it will reduce the difficulties later in the Design and Implementation Phases.

Once agreement has been reached on the System Definition, it then formally becomes the specification for the following important but difficult Design Phase.

5
STEP 4: DESIGN PHASE

The main objective of this phase is to obtain a detailed system design for implementation. This means that the entire system has to be defined in terms of information flow, files, volumes, print designs, procedures, forms, program specifications, etc. In addition to these, a revised estimate of the operational cost of the system is calculated after the design has been completed and a revised plan for implementation produced.

The main activities which are undertaken in this phase are:

- Finalise information flow, data elements, output requirements, data relationships, etc.
- Identify master files, working files, data volumes, frequency of updating, length of retention, speed of response required from files.
- Determine the file organisation and layouts.
- Specify input layouts, frequency, etc.
- Define reporting requirements, volume, frequency, distribution.
- Develop overall system logic.
- Determine controls and audit procedures.
- Identify computer programs and manual procedures required.
- Prepare program specifications.
- Develop general test requirements (type of data, source, control checks, etc.).
- Revise estimate of operational costs of system.
- Produce detailed plan of implementation.
- Document the Design Phase in a report for user and systems management.
- Decide which storage devices should be used.
- Decide on division of the computer-based parts of the system into individual computer runs.

The broad information flow will have been identified in the system definition phase. In the design phase this should be reviewed and broken down into a greater level of detail. A master flow chart of the system is then produced, showing all the inputs, files, processing stages, error recycling and outputs of the system. Care must be taken to identify the different time factors, for example, frequency of reports, length of retention of files and input data, etc.

STEP 4: DESIGN PHASE

An important tool for use at this stage is decision tables. They are invaluable in defining the relationships between data, programs and users. Decision tables have a range of uses, as described in Part II, 11, and are particularly effective at the system design stage. In conjunction with flowcharting, they are major means of developing a systems design.

It is at this stage—as an integral part of the design process—that systems and audit controls should be built into the system. This implies consultations with company auditors. Many systems people make the mistake of not paying sufficient attention to controls at this stage, beyond normal validation of fields and simple check totals. Controls are usually very difficult to introduce at a later date. They are discussed in depth in Part III, 15.

An important task for the systems analyst at this stage is the identification of computer programs and the development of program specifications. Criteria for dividing the processing procedure into programs are:

- Size of each processing step (the installation should have advisory standards for the maximum and prefered size of program, depending on the computer and configuration).
- Logical units (input validation, printing runs, sorts, etc., are clearly identifiable as performing basically different functions).
- Ease of amendment (the steps can be classified according to their likelihood of change; the areas of probable stability can then be separated from the ones of probable change).
- Time relationships (there will be natural breaks in processing runs due to waiting for user action, etc.).
- Suitability for controlling (discussions with the auditors will pinpoint certain stages and check points).
- Suitability for testing (it is important to bear in mind implementation factors at this design stage; the possibilities for testing programs and portions of the system in terms of data source, volumes, etc., should be considered).

A very important strategy to follow in dividing the system into units is that of *modularity*. This is a basic philosophy applicable to any kind of problem solving: divide the overall problem into sub-divisions which can be more easily handled separately. The implications, however, go much further. For example:

- The job is separated into skilled and less skilled parts, so the personnel resources available can be used to best advantage.
- Modular systems make documentation easier to create.
- Modification is *much* easier.
- The system is easier to test when it has a modular structure.
- The design of modules forces a thorough analysis and understanding of the problem.

- Integration with other systems is easier with modular systems.
- General-purpose modules (for example, input control) can be used in other systems.

Modular systems design is not easy. It takes skill and experience, although the skill is rapidly acquired by a reasonably experienced systems analyst.

Another example of flexibility in systems design is the use of parameters. The principle here is that a general-purpose module is designed which can be tailored to suit a number of different purposes within a given range. The tailoring is accomplished by providing the appropriate data to the system before proper running starts, so that it is then 'set up' for the particular run in question. A simple example of this is the retrieval of information from a file. Let us assume that the file contains the following data elements:

- Customer number.
- Name and address.
- Category of customer.
- Discount rate.
- Purchases this year to date.
- Total purchases last year.
- Date at which first became customer.
- Sales region.

This file contains information which might be used by sales accounting staff, sales offices and marketing staff. Their needs are likely to be different and also to change over time. For example, market research personnel may at present be investigating all customers of a given category, but the emphasis may change to restrict their analysis to customers of a certain minimum sales value within that category. If a special program has been written to cover their requirement, it will cause a rewrite of the program to produce a report for their second requirement. This is easily catered for, however, by a parameter approach. A generalised program (or suite of programs) is designed which will print out all combinations of data elements from the file. Parameter input is designed to allow selection of the data elements, as shown in the sample form, Fig. 5.1.

If the user marks the box for 'ALL', this means that all customers meeting the specification should be reported on. If he is interested only in certain specific customers, however, he has the possibility to specify these on subsidiary sheets by marking 'FOLLOWING' and using the form shown in Fig. 5.2.

Similarly, customer categories, discount rates and sales regions may be specified. These forms are then punched into cards or tape and act as input to the generalised program, which searches the file record by record, selecting out the appropriate customers according to the logical 'AND'

STEP 4: DESIGN PHASE

relationship indicated on the form. The general scheme of this processing can be represented by the diagram in Fig. 5.3.

Note that the key program is the compilation program which sets up the generalised program for a specific purpose. Depending on the file organisation, it may not be necessary to sort the file and this can also be set up and run by means of the parameter which act as sort keys.

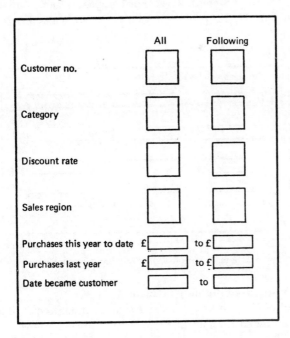

Fig. 5.1. Sample form for specifying input parameters—I.

The authors have used this design approach on many occasions, when information requirements based on a range of data elements were uncertain. The benefits are flexibility for the user and the avoidance of designing special retrieval runs for each new request for the analyst. Whilst nothing can replace the spark of human creativity which is always necessary in the design phase, there are a number of inputs which can actively help the systems analyst in this stage:

(1) *Examine Other Systems Used in Similar Situations*

It is very rare that a truly new problem arises. Somebody else will always have had to face essentially the same problem and, while they may have failed to solve it, their experience is bound to be useful if only to point out the way *not* to do it. So constant review of the literature is called for, plus widespread enquiries about other systems in the same general field.

(2) *Make Use of Available Techniques*

There are plenty of aids to help the analyst in design work. For example, decision tables are a fine means for defining relationships between data, activities, etc. (see Part II, 11). Then there is of course flowcharting (discussed in Part II, 10). Statistical techniques, too, are always a valuable tool for the systems analyst. For complex systems, simulation (discussed in Part II, 12) is often the only way to evaluate alternatives in design.

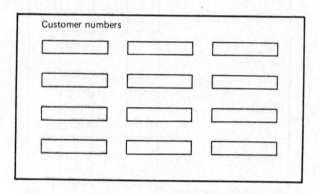

Fig. 5.2. Sample form for specifying input parameters—II.

(3) *Use Group 'Brainstorming' Sessions*

It is a bad practice to leave all the design creation process to one person or even a limited number of persons. It is amazing how ideas can be generated by the interaction of minds. The policy here should be to keep the group small (it may be necessary to have several group sessions with different persons) and to have virtually no restriction on ideas. Most of the suggestions will probably be unusable, but it is also likely that *some* of them will be good.

The systems design phase culminates in a document which is then approved or modified by user and ADP management. The whole project may be dropped if it is impossible to achieve a system design which satisfies both. In the more likely event, however, that it is accepted, after modification, this document acts as the specification for the implementation phase.

Before proceeding with further work the analyst should ensure that the design specification is not only passively accepted but formally 'signed over' as the specification for the implementation phase. Failure to do this will frequently lead to requests from user departments for alterations to the system design after the analyst has considered it 'frozen'. This will inevit-

STEP 4: DESIGN PHASE

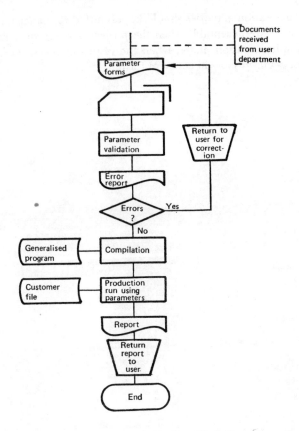

Fig. 5.3. Scheme of processing for generalised file interrogation using parameter input.

ably lead to delays and to dissatisfaction for all parties. Furthermore, the formal acceptance of the project is restatement of user commitment to the project.

The contents of design specification are:

- Narrative of system functions.
- General flowchart showing major system functions (both manual and computer).
- Identification and relationship of input, output and files.
- Identification and relationship of programs.
- Detailed flowcharts showing interaction of input, output and files.
- Program specifications.
- File specifications.

In addition to the design specification, a revised estimate of project costs and implementation activities should be submitted to management at this stage. If these are acceptable, then the project is ready to move into the exciting implementation phase, when the results of the work invested in the project begin to show in concrete form.

6
STEP 5 : IMPLEMENTATION PHASE

The objective of this phase is to obtain an operational system, fully documented. The main activities which take place are :

- Write and debug all computer programs.
- Create master files.
- Prepare documentation for ADP and user departments.
- Acquire all necessary equipment, stationery, etc.
- Train ADP and user staff to use system.
- Supervise phasing-in of system parts.
- The testing and proving of all parts of the system.

The implementation phase is where many well-designed systems go wrong. They either simply fail to get off the ground or are implemented without proper control so that the master files are thoroughly unreliable, programs go operational while full of errors and users do not know how to work with the system. If this happens, the project is lost, because the resulting lack of confidence in the system will permanently bias the user against it. Furthermore, other systems to be installed later will suffer from doubtful cooperation by users.

The analyst must retain tight control during the implementation phase. A professional approach to project control should be followed. The main points to follow are :

- Use a formal planning and control technique. A characteristic of systems work is the amount of change which takes place during a project. Without applying a flexible method of planning and control, this change can lead to ineffective work and even confusion. A plan that is changed is not necessarily an indication of bad planning; it is more likely to reflect the original flexibility of its construction. Suitable techniques for planning and control are described in Part IV, 21.
- Set objectives for each person involved in the project. It is not sufficient to rely just on the overall project objective. Each individual should know what is expected of him. This is sound management theory applicable to any type of project, but is particularly important in the case of systems work, because there are so many

vaguely defined activities, such as establishing opinions, making subjective assessments, interviewing, etc. Such activities may stretch out in time very easily without any significant contribution to the project being achieved. The project leader must therefore decide for each member what is a realistic time for completion of the activity in order to achieve a given result.
- Ensure that all phases of work are carried out in accordance with predetermined standards. Part III, 16 discusses the need for standards and identifies the main areas where they should be applied.

Note that it is the systems analyst's responsibility to train both the user and data processing staff in the use of the new system. This includes preparation of input, error correction, computer operating, job scheduling, interpretation of output, choice of stationery, application of check totals, distribution of reports, time requirements, etc. A real 'sales job' should be done, if necessary making use of classroom presentations and even films if the system is big enough to warrant it. User and operating manuals should be distributed in good time.

The system should not go operational until it has been proven error-free and the users are familiar with its operation. This implies thorough testing. Guidelines for system testing are:

- Test each system module separately.
- Use live data as well as test data.
- Do a parallel run before going solo.
- Test the system as a unit.
- Get the user to test the system.
- Deliberately put data through from a totally different system, to make sure that it all gets rejected.
- Try month-beginning, month-end, year-beginning and year-end procedures.
- Get the company auditors to test the system.

In addition to the above, the systems analyst should go 'on shift' *with* the users for the first few days or longer until the system is thoroughly run in. Apart from being on the spot in the event of anything going wrong, it is very good psychology as well as practical support for the persons using the system for the first time, who may be uncertain or even nervous in using it.

7
STEP 6 : EVALUATION PHASE

The work of a systems analyst on any given project does not cease when the system becomes operational and error-free. Unfortunately, too many systems analysts do consider their work complete at this stage and move on to new projects without ever carrying out the very necessary evaluation phase.

The evaluation phase takes place after the system has been running error-free for several cycles. For example, in a system which is run once a month it is best to wait five to six months after implementation before carrying out the post-evaluation. This waiting period is necessary to make sure that the users and ADP staff are thoroughly familiar with the system. A weekly system should run for two months before post-evaluation and a daily system three to four weeks.

The purpose of the evaluation of the operational system is threefold. Firstly, the efficiency of the system is examined to see where improvements can be made. Secondly, the achievements of the system are compared to the objectives originally set. Thirdly, the evaluation provides valuable feedback to the systems analyst, so that he may learn from the good and the bad points of the system.

To carry out the post-evaluation, the analyst should assess the performance of the system from the following aspects :

- Actual cost.
- Realised benefits.
- Timing.
- User satisfaction.
- Error rates.
- Problem areas.

These should be compared with those foreseen at the Feasibility and Design Stages. Significant variances should be investigated to determine the cause. In this way areas for improvement can be identified. It is frequently possible to attain substantial improvements by making relatively small changes. For example, modification of one program may drastically reduce computer processing time. Data handling and input, too, can often be improved significantly once the system is running.

The findings of the evaluation phase should be documented in a report for user and ADP management, summarizing the achievements of the system with recommendations for improvement. It is sometimes the case that benefits expected are in fact never realised and the decision may have to be made to discontinue the system and reduce losses.

After the initial post-implementation evaluation, the system should be reviewed annually. Most systems constantly undergo amendments because of the changing nature of business and very often so many amendments are made that the system is no longer meeting its original purpose. To avoid having systems which, although apparently running efficiently, are obsolete, it is therefore an important part of systems analysis to review every operational system periodically to see whether or not it is still meeting its objectives, or whether or not its original objectives are still valid.

8
REVIEW OF THE SIX STEPS OF SYSTEMS ANALYSIS

Although we have considered each of the steps separately, they are, of course, related to each other by the dependence of each step on the successful completion of the preceding one. Furthermore, the steps have been designed to provide checkpoints where decisions can be made as to whether or not it is worthwhile to continue the project.

Their relationship and the way in which the checkpoints operate can be seen from Fig. 8.1.

The checkpoints ensure that the project does not become unnecessarily advanced before realising that the system cannot after all meet the requirements. Lack of checkpoints is a common weakness in the work of most systems analysts. It is very important to formalize into specific decision points the build-up of information on the true feasibility of the proposed system.

An additional point to notice in reviewing the six steps of systems analysis is that none of the main steps is devoted to a 'Study of the Existing System', which is beloved by specialists in the traditional Organisation and Methods approach. Of course it is absolutely essential to study the existing system, but with computers providing wider and wider scope for the design of new systems it is important not to orientate oneself too much to the existing system. The approach throughout should be aggressive—what is the best way of accomplishing this objective, regardless of the means presently used? This contrasts with the usual approach of : —how can I improve this existing system? Most business systems are the result of a continuous process of evolution and a system at any one time has therefore of necessity largely been determined by past requirements rather than current ones. There are usually many useless and often troublesome relics inherited from bygone days, rather like the appendix in the human being.

The authors regularly encounter systems in companies of all sizes and reputation which are running on modern, fast, expensive computer systems, but are really old procedures which were first converted to punch card systems running on traditional card equipment. These in turn were converted to a computer system based on punch cards, then to magnetic tape and disks operating in a sophisticated mode with multi-programming and data communications links, etc.

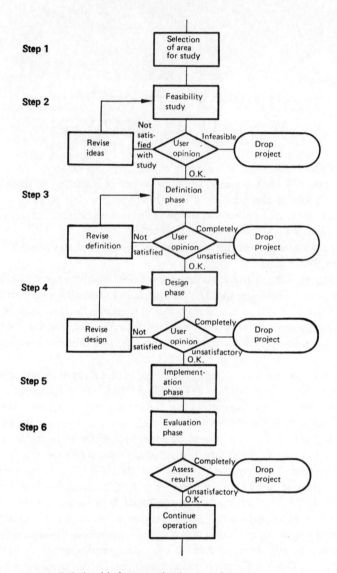

Fig. 8.1. Relationship between the six steps of systems analysis and the identification of project checkpoints.

One is reminded of the analogy of the automatic dish washer. Before modern electric dish washers came on to the market for everyone to see, it was puzzling to the uninitiated to visualize how they would work. Most people thought in terms of two mechanical arms which moved out of the kitchen wall, one of them grasping each plate in turn while the other scrubbed it. This was because people found it difficult to visualize a solution different from the traditional way of doing the job. In the event, the new and equally effective method proved to be very much simpler than a mechanical arm system of the popular vision!

There are, of course, implications in the approach advocated here which do not arise with the traditional approaches depending heavily on the existing system. The traditional approach has the advantage in that the improved system is almost certainly going to be based on the same personnel, organisation and company policy as before. No drastic change is usually necessary to implement the new system. In the approach described in this book, however, a thoroughly fresh look at the true requirements of the user may result in a system radically different from the old one and may frequently involve organisational changes, for example. This is one reason why it is highly desirable that the systems department should be a staff function serving the senior management of the company and not reporting to any one functional area such as accounting.

The existing system is an essential base from which to obtain information about the requirements of the company and provides a yardstick for each of the six steps. The emphasis throughout, however, should be on defining current requirements and not on finding new ways to meet old requirements.

PART II
Techniques

9
FACT GATHERING

Some experts consider that a systems analyst's job is approximately 40% technical and 60% human relations. This 60% human relations content does itself have a number of facets, among the most important of which may be considered :

- Choosing a team to conduct a project.
- Leading a team.
- Communicating ideas.
- Giving technical instructions.
- Obtaining information.

This last category of obtaining information itself covers a number of techniques, but easily the most important of these is interviewing. It is, after all, the principal way in which a systems analyst obtains the detailed information he needs about the present system and the requirements for the new system. Notwithstanding its importance, this part of a systems analyst's training rarely receives the attention it deserves. In cases where systems analysts are recruited as such, and due consideration is paid to the personality requirements of the job, the results may not be too serious, but in cases where systems analysts are recruited for their technical knowledge of computers alone (as is often the case with ex-programmers) the results may be little short of disastrous. Indeed, it is the authors' opinion that much of the present gap, which undoubtedly exists between the technical capability of computers and the realisation of that potential, is due to the neglect of the human relations aspects of a systems analyst's task.

How, therefore, may this part of a systems analyst's training be undertaken? There are, at present, few courses which even include any mention of the interviewing technique, it being supposed, presumably, that obtaining information from other individuals is not of sufficient importance to justify the teaching of the techniques required, or alternatively that the ability to conduct a successful interview is inherent and not one that can be taught.

It is to a certain extent true that human relations skills require some natural aptitude, but it is not right to conclude that no improvement to whatever natural aptitude the individual possesses can be made by careful training.

The remainder of this chapter lists the basic principles of a sound interviewing technique which are universally applicable, some of the other ways of obtaining information and then illustrates an unsatisfactory and satisfactory way to conduct a typical interview. It has been found that the critique of simulated interviews conducted under the supervision of an experienced analyst is a particularly valuable way to improve the interviewing technique of inexperienced staff, and the examples in this chapter serve to extend this technique to private study. They may also be used to provide a script for use in either informal 'in-house' or more organised training schemes.

9.1 Interviewing

Here are 14 guidelines for successful interviews :

9.1.1 *Make an appointment.* Whenever a lengthy interview (more than 15 minutes) is planned or when making a new contact, it is highly advisable to make an appointment. This serves two major purposes. Firstly it ensures that the interviewee is available and thus minimizes the possible waste of time, and secondly, if advance notice is given of the subject of the interview, it will enable the interviewee to collect the necessary information.

9.1.2 *Be prepared.* The interviewer must take great care to be prepared. He must be perfectly clear as to what he wishes to obtain from the interview and whether this is fact or opinion. It is recommended that a short check-list be drawn up to ensure that no points are missed. This will also have the effect of minimizing the disturbance to the interviewee through having to go back to him later to obtain supplementary information.

9.1.3 *State the reason for the interview.* Whether or not the interviewee has previously been acquainted with the purpose of the meeting, it is as well to prevent any possible misunderstanding by stating briefly the purpose at the start of the interview. Some such sentence as : 'I would like to obtain some details about the preparation of the weekly timesheets' will usually be sufficient.

9.1.4 *Observe good manners.* This heading can also be taken to include some of the other points on this list, notably 9.1.1 and 9.1.6, which may be regarded as particular examples of good manners. It is included as a separate heading, however, to remind the interviewer of the necessity to observe normal good manners throughout the interviews. Particular examples of ill-manners frequently occurring in business circles which may cause offence are :

- Smoking without asking permission.
- Failure to introduce himself on the part of the interviewer.

- Not paying attention to the speaker.
- Unpunctuality.

9.1.5 *Use interviewee's language.* Most departments within any organisation have their own specialised jargon. Failure to understand this language may mean that the systems analyst will not obtain all the information he requires or—worse still—may misinterpret it. In any case, it may well leave the interviewee with the impression that the computer department does not understand his problems. It is, therefore, extremely desirable that the systems analyst, as part of his preparation, makes sure he understands the departmental jargon.

9.1.6 *Don't interrupt.* The systems analyst must guide the interview to a considerable extent. It is he after all who requires the information. He must not, however, interrupt the interviewee if this can possibly be avoided. It is far better to wait until a convenient moment occurs to turn the interview back to a more pertinent line of enquiry than to interrupt, as this may well inhibit the interviewee from expressing himself freely later on.

9.1.7 *Avoid Yes/No questions.* The object of an interview is normally to find out facts or opinions from the interviewee; this being so, it is more effective to avoid questions which may be answered 'yes' or 'no'. At first sight these questions may appear the easiest and quickest way of obtaining facts, but in practice they will invariably mean that important qualifying information is omitted. In any case, the subject will not feel he has been really consulted. Any system where the design is based on yes/no information at the fact-gathering stage is liable to encounter difficulties during implementation.

9.1.8 *Don't express your own opinion.* When conducting an interview a systems analyst will often be asked to state his opinion about persons, departments or systems. It is dangerous to express such opinions and the temptation to do so must be resisted. An extension of this guideline, which is perhaps even more important, is that the systems analyst must not argue with the interviewee. Except for rare occasions, his duty is to obtain facts and opinions; this purpose will not be served by arguing!

9.1.9 *Compliment when possible.* Most people are encouraged by being paid an occasional compliment, especially when it shows the systems analyst is genuinely interested in the person to whom he is talking. Even such a general remark as 'That's a fine plant you have on your windowsill, Mr Brown' is very effective in encouraging the subject of the interview to express himself freely.

9.1.10 *Distinguish between fact and opinion.* Many people express their own (or their department's) opinions as proven facts. The systems analyst must take special care to ascertain which of the information he has been given is fact and which is opinion.

9.1.11 *Distinguish between need and desire.* This is particularly important when assessing the design requirements for a new system. The general remarks in the preceding paragraph apply to the necessity to distinguish between need and desire.

9.1.12 *Don't be asked to leave.* In many ways the success of an interview may be judged on the final impression left on the interviewee. If the systems analyst has failed to observe the signs that the interview has gone on long enough (or has exceeded the time limit if this was previously set), the final impression will be adverse. Observe, therefore, signs of restlessness, particularly glancing at a watch or clock, and be prepared to continue later if the purpose of the visit has not been fully met.

9.1.13 *Check information obtained.* With the best will in the world people often make mistakes when quoting facts or alternatively may give only partial or even obsolete information. It is, therefore, vital that the systems analyst should *cross-check facts whenever possible*.

9.1.14 *Confirm results of discussion.* To prevent any future disagreement it is strongly advised that facts obtained or agreement reached during a meeting are confirmed afterwards in writing.

One point that often worries people about interviews is whether it is permissible to take notes. Where detailed facts are concerned this is really vital, but even in more general discussions it is much safer to make a note of the essential facts at the time rather than to rely on memory later. It is not necessary for a verbatim report to be made—indeed, this would inhibit most interviewees—but a series of short notes will prove invaluable when assessing the results of the interview later. Of course, before taking notes the permission of the interviewee should be sought and their purpose explained in order to prevent the thought that 'He's writing it all down to quote against me later'.

9.2 Other Fact Finding Techniques

Although interviewing is the most important fact-finding technique, it is not the only one, and consideration must be given to the other sources of information available to a systems analyst. These will include examination of company documentation both within the data processing and user departments (and sometimes other departments as well), examination of trade or national statistics, the use of random sampling techniques, observation and the use of questionnaires.

9.2.1 *Company documentation.* Searches of company documentation will, of course, cover a multitude of sources of information ranging from minutes of board meetings (which may reveal areas of future expansion and thus influence file sizes, etc.) to the detailed documentation of an existing data processing system. The documents searched in the analysis and design phases of an assignment will, of course, vary from system to system and company to company. A few general points should be remembered in all cases.

Firstly, check to ensure that the information is still current. It is of no value to base system design on obsolescent information.

Secondly, cross-check information wherever possible. There is a tendency to assume that written information is more reliable than verbal information. This is not necessarily true. The mere fact that the information has been committed to writing does not guarantee its accuracy.

Thirdly, ensure the information is relevant. It is easy to be misled into making the basic mistake of accepting statistics which are *almost* what are wanted as being *exactly* what is wanted. If, for example, in designing a computer system the analyst requires to know the number of weekly paid employees who are paid by direct bank transfer, it is not sufficient to be told the percentage of all employees who are paid in this way. The information will not be correct for his purposes.

9.2.2 *Examination of trade or national statistics.* All the points listed above are equally important when seeking facts from trade or national statistics, but special care must be taken in assessing the relevance of the information to the needs of the study being performed.

9.2.3 *Random sampling techniques.* Random sampling techniques are an extremely valuable aid during the fact-finding phase of an assignment, but great care must be taken if the results are to be properly applicable to the whole population of the study. Basically the technique consists of selecting a proportion (the sample) of the total (the population) and, based on the assumption that the sample is representative of the population, drawing inferences about the whole, for example, if a sample of 10% of the invoices received in a week had an average value of £20 and a total value of £4,000 it could be inferred that the average value of all invoices was £20 and their total value was £40,000. In the use of this technique there is, however, a major pitfall. If the size of the sample is insufficiently large to enable any statistically valid results to be drawn from its analysis, the sample chosen is not representative of the whole. In the above example the invoices from some major suppliers might only be received monthly, thus distorting the picture obtained from analysis of any one week's invoices. This pitfall is explained in any textbook on statistical techniques and the analyst is strongly advised to study one of these or consult an experienced practitioner before using the technique himself.

9.2.4 *Observation.* One of the ways in which an experienced analyst collects facts is by observation. This is not a specific technique which can be taught but is rather a subjective assessment of those factors likely to prove of importance. Observation may take the form of simply 'keeping one's eyes open', by cultivating a general interest in all procedures. But in its more specific form it implies formally observing an activity to identify problems, skill characteristics, time needed, environmental constraints, arrival of information, etc. The key to successful observation is to take a purely passive and patient role, not interfering in any way with the process itself, which must be allowed to take place as it normally would. The use of observation as a fact-finding aid will come readily to the analyst trained in O & M, work study and similar techniques, but perhaps rather less readily to the ex-programmer.

9.2.5 *The use of questionnaires.* Facts obtained from answers to questionnaires are less likely to be accurate than those obtained from answers to questions posed at a direct interview. This is because the interviewer is in a position to explain his questions more clearly and to interpret the replies in the light of the overall assignment. He is also able to make a subjective estimate of the probable validity of the information. All these aids are denied to the analyst working with information supplied in answer to a questionnaire. He does not even know whether the person who completed it made any attempt to do so accurately or merely guessed. For these reasons alone the use of questionnaires should be avoided where possible and personal interviews used instead.

There is also a further reason why questionnaires often result in highly suspect information. This is because, unless designed by an expert, the questions asked may prove misleading or ambiguous. For example, a questionnaire on staff turnover included the question 'What was the percentage staff turnover in the ADP department in 1970?' No further instructions were given, so answers could have been calculated on the staff level at the beginning, middle or end of the year and may have (or may not) included temporary staff. Results obtained from questionnaires of this nature are meaningless.

There are, of course, some instances, e.g. where the size of the survey to be undertaken is too great for individual interviews, that the use of questionnaires is the only possible method. In these cases great care must be taken in framing the questions and, if possible, expert advice obtained. The use of semi-trained personnel to present the questionnaire and assist the recipient in completing it may well be worthwhile to ensure a high response rate and reduce anomalies. Furthermore, the use of a well-worded letter explaining the need for the survey and requesting the assistance of the person to be questioned will also serve to improve the accuracy of the replies.

9.2.6 *Recording facts.* Mention has already been made of the value of taking notes during an interview. This is also true of any other fact-finding technique. *Facts should be recorded as soon as possible.* When obtaining information about an existing system a useful *pro forma* is shown in Part III, 16. This enables the analyst to note directly in either narrative or chart form the steps in a procedure in a format that enables him quickly to construct a flowchart or full narrative description and also provides a suitable system description to refer back to users for verification. By following a logical step-by-step progression, it also reduces the possibility of subsequently having to check back to complete the description.

Play-acted Interview
(How not to do it)

A = Systems analyst
B = Accounts manager being interviewed

A Good morning, I'm from the computer department. We sent you a memo telling you I was coming at this time.
B Yes, I did agree with Mr Poppit, manager of Management Service Department, that I would be free now for discussion with one of his personnel.
A Well, I'm the one doing the job. You see, we want to computerise your department. You have an awful lot of staff in your department and we feel that we can replace a lot of them by the computer.
B What did you say? Take away some of my staff? My staff are badly overworked as it is.
A No, you don't understand. Let me explain. We are going to set up a master file of all customer accounts and keep them on disk. Of course, we'll do an updating run before we use this file to input to the invoicing run and monthly sort and merge. Here, let me show you on this flowchart. (Shows flowchart, explaining it in detail.) (B looks at his watch.)
B This is all very well, but you know my department is probably the most important one in the company. We control all financial matters very carefully and we . . .
A (Interrupting)
   Oh, don't worry about that! The computer is much more accurate than your staff can be. (Lights cigarette.) In fact, that's going to be a problem after the system has been installed. Unfortunately, we'll still have to rely to some extent on human beings. I mean your people will still have to provide the input data to the system. They'll have to improve a lot, though.

B   I'd like you to know that my department won the group award last year for courtesy and efficiency. Furthermore, we always . . .
A   (Interrupting) Yes, but we are talking here about a computer, which will do much better than your staff. Just wait and see.
B   No, I will *not* wait and see. (Looks at his watch.) I am going to speak to Mr Poppit personally about this matter and we'll see then if . . .
A   (Interrupting) You can if you like, but I'm afraid it's too late. I've already put in my recommendations and I'm pretty sure they'll be accepted.
B   You've already put in your recommendations! Why did you come to see me then, may I ask? I shall see Sir John himself about this. Now I'm afraid I must go to a meeting, so if you'll excuse me I'll . . .
A   (Interrupting) Well, I just have some more questions to put to you We need more data about your activities. When do you require the statements of customer accounts? Every month?
B   I think we can discuss this some other time. I'm in a hurry now.
A   Well, just one last question. How many customers are there who . . .
B   (Interrupting) Good-day, Mr . . . What is your name? My secretary will show you the way out!

Play-acted Interview
(How to do it)

A = Systems analyst
B = Accounts manager being interviewed

A   Good morning, Mr Smith. My name is Dolittle from the Management Services Department.
B   Ah, good morning, Mr Dolittle.
A   It's very kind of you to offer to give up more of your time like this. We realise you're a very busy man and I'll try not to take up too much of your time.
B   Yes, I must admit I am rather busy these days. What can I do for you?
A   Well, as you know, we in the Management Services Department have been trying to see how we can improve the support we give for managers. We have already done some preliminary work and, in fact, you kindly let us talk to some of your staff last month. We have tried to concentrate on your area to start with because your department is a key part of the company, which affects most other company activities. So naturally we want to take advantage of your experience and ideas.
B   Well, I'll certainly try and assist you all I can.

A Thank you. One of the problems with which we are very much concerned at the moment is the enormous amount of paper-work which is being thrust on departments such as yours. You have probably felt this problem increasing over the years.

B You are certainly right! It's getting quite out of hand. Why, I myself spend half my time just filling in forms and writing reports!

A Quite, and this is what is worrying us. Managers like yourselves want to concentrate on true management activities like decision making, control of the business, etc. And your staff could be doing the work you really would like them to do, that is carrying out the functional control you desire.

B Exactly so, I must admit that most of my staff are doing nothing more than clerical tasks, though they're all of them good people.

A Well, we've been trying to get on top of this problem and we have some proposals on which we'd very much appreciate your advice. We think one way in which we could help alleviate your problem is to push away the routine burdensome paperwork on to a machine, the computer. I'm afraid the computer is a very primitive and stupid machine and really all it can do is obey what we tell it to do, but for this reason it is useful for carrying out the donkey work. In this way, you and your staff will be left with more time for decision making and control of the business.

B Well, I am certainly interested in what you are trying to do and will do my best to give you what information you require. I am glad that at long last somebody is trying to do what I have been saying should be done for the last ten years.

A If we may, we'll contact you further in the next week or so to ask your permission to talk to some more of your staff. You've been most helpful today indeed. I'll be confirming the details we've discussed in a memo to you. May I thank you very much indeed for giving up so much of your time.

B It was a pleasure. Good-day, Mr Dolittle.

A Good-day, Mr Smith.

10
FLOWCHARTING

Flowcharting is a most useful technique for the systems analyst and is very widely used. It is perhaps because of this universal usage that the technique varies so much between various users, even, in many instances, between different individuals within the same company.

There is a tendency to regard flowcharting as a technique unique to data processing, but the basic concept of a diagrammatic representation showing also a time sequence far precedes the advent of computers. This technique has proved particularly useful, however, in this field and data processing is now probably the most common application.

Efforts are being made to standardise the technique not only within user organisations but also between users and more recently between users of various manufacturers' equipment. This last point is of particular importance since the use of the technique has been fostered by computer manufacturers who have all adopted different symbols and conventions. The symbols and examples used in this book follow the International Standards Organisation (I.S.O.) recommendation (ref. R. 1028).

10.1 The Technique

In all cases flowcharting is the diagrammatic representation of information expressed in a time sequence. In a data processing environment the symbols recommended by the I.S.O. are those given in Fig. 10.1.

In addition to the basic symbols there are a number of conventions to be observed in drawing any data processing flowchart. The major conventions are:

(1) The sequence of the flowchart is (unless otherwise stated) down the page and from left to right.
(2) The sequence is shown by lines joining the symbols which themselves represent operations performed.
(3) When drawing a chart on more than one sheet of paper the connectors joining different pages (or parts of a chart within one page) must be adequately referenced.

The full list of conventions relating to the drawing of flowcharts is shown in Fig. 10.2.

10.2 Use of the Technique

Even within the data processing environment there are different uses of the technique, namely :

- Illustrating the logic of a program.
- Use in illustrating systems (either in total or part).
- Illustrating the interaction of systems.

The systems use of flowcharting can be further sub-divided as an aid :

(1) To understanding a present system.
(2) To designing or evaluating a new system.
(3) To illustrate a proposed system for the information of other persons (programmers, operators, and, sometimes, users and management).

To meet these differing requirements, systems flowcharts are drawn in varying degrees of detail. The example in Fig. 10.3 shows part of a flowchart drawn by a systems analyst to assist in the understanding of an existing system.

The discipline necessary for drawing a flowchart of this type enforces a thorough understanding of the system and quickly reveals any gaps in knowledge by a breakdown in the continuity of the chart.

In the design phase of a new system a flowchart can be extremely useful in revealing unnecessary processing or a break in continuity. Initially it is not normally desirable to draw a flowchart in full detail but rather to illustrate the input and output phases of the system and frequently the clerical work associated with them. A chart of this nature (shown in Fig. 10.4) may also be useful as supplementary evidence in presenting a proposed system to management.

To illustrate a proposed system to other staff within the data processing department it will normally be necessary to prepare a flowchart which illustrates the departmental activities in detail whilst the user activity is shown only in outline. An example of a systems flowchart prepared for illustrating a system to data processing personnel (in this case for inclusion in the operating instructions) is shown in Fig. 10.5. Depending on the division of responsibilities between analysts and programmers within the organisation, it may also be necessary for the analyst to draw program flowcharts.

10.3 Drawing a Flowchart

The first step in drawing any flowchart is to be quite clear about what area of activity is to be shown and for what purpose the flowchart is to

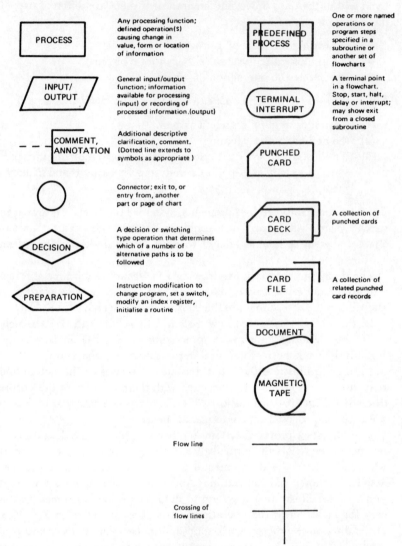

Fig. 10.1. Flowcharting symbols of the International Standards Organisation.

FLOWCHARTING

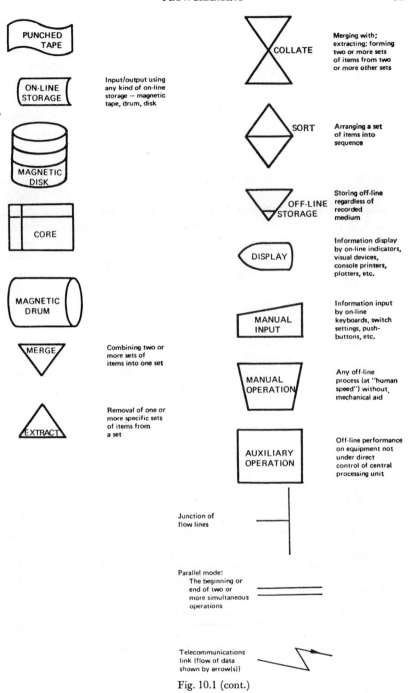

Fig. 10.1 (cont.)

be drawn. These objectives having been clearly identified, the second stage is to obtain the information to be included. Depending on the complexity of the activities to be charted, it may be necessary to produce a simple narrative description as an interim stage. A suitable *pro forma* for this purpose is shown in Fig. 16.1 in Part III, 16 and its use explained on page 49.

Once this information has been obtained, the way is clear for the actual charting to begin. It is advisable to use a specially designed form for this purpose as this will reduce the actual amount of drawing to be done and also help to ensure that a neat and presentable result is produced at the first attempt. When choosing such a form, however, the constraints imposed by the documentation standards should be borne in mind (see Part III, 16). The forms supplied for this purpose by various manufacturers do not always meet the requirements of good documentation and it may well be more desirable for an 'in-house' form to be designed for this purpose. A form of this type which meets all the requirements of good documentation outlined in Part III, 16 is shown in Fig. 15.3.

When drawing a flowchart the following points should always be considered:

(1) Do not hesitate to specify detailed procedures outside the body of the main flowchart by the use of the predefined process symbol. This has the effect of simplifying the main flowchart and thus allows the main points to be assimilated more readily.

(2) Make full use of the comment or annotation symbol to ensure that the chart is comprehensible.

---

1. The general direction of flow in any flowchart is:
   From top to bottom
   From left to right.
2. Arrowheads will be used whenever the flow of information *or sequence of events* is not as indicated in 1 above *and* whenever necessary to improve the clarity of the flowchart.
3. Flowlines crossing shall imply no logical connection between those lines.
4. Two or more flowlines may join without any explanatory note. Two or more flowlines may only diverge at a symbol which is annotated to show under what conditions divergence occurs.
5. Flowchart symbols may be drawn any size, but the ratio between dimensions should be maintained within reasonable limits in order to facilitate recognition.
6. Any flowchart should be identified with a title, date and the name of the author.
7. Annotation and cross references should be made when the meaning is not apparent from the symbol(s) used.

Fig. 10.2. Conventions for the use of flowcharting in data processing

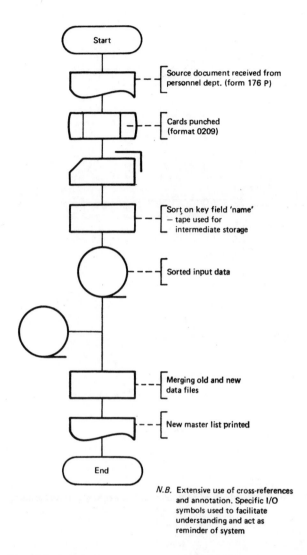

Fig. 10.3. System flowchart for a system to update a personnel record file (chart drawn to facilitate understanding of present system).

(3) Be as neat and tidy in drawing as possible. A little more care exercised in the preparation of a chart can eliminate the need to re-draw and thus lead to an overall saving in time and effort.
When the chart is complete it must always :
(a) Be fully identified. }
(b) Be dated.   }   see Part III, 16.

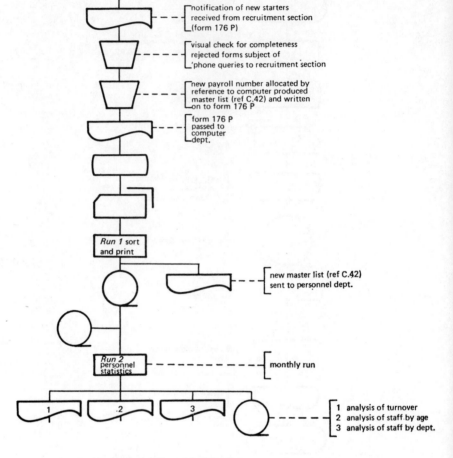

Fig. 10.4. Flowchart of part of a personnel statistics procedure (chart drawn with emphasis on clerical procedures).

(c) Have the author's name on it.
(d) Be checked to ensure that no uncompleted flow remains.

10.4 Automated Flowcharting

In addition to the manually produced flowcharts described earlier in this chapter, it is also possible to produce flowcharts by computer. The principal uses of computer generated flowcharts are:

(1) To provide up-to-date flowcharts of programs for documentation purposes.

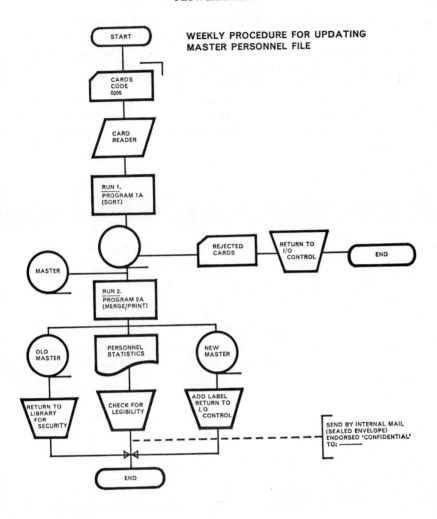

Fig. 10.5. Flowchart of personnel statistics procedure (for inclusion in computer operations documentation).

(2) To provide a programmer with a flowchart for use in debugging a program.

For the first of these uses the computer generated flowchart has a number of advantages, principally because it produces flowcharts as the program actually is—not as the programmer intended to write it. Secondly, it enables the programmer to proceed immediately after testing with further programming without neglect of his documentation responsibilities.

As a debugging aid most of the software packages currently available provide, in addition to a fully detailed flowchart, details of incomplete program loops, branches to non-existent labels, etc. The exact facilities available depend on the package being used. All these packages work in basically the same way, providing output based on a program source deck, and, therefore, cannot be used until after the program has been written. For this reason they cannot fully replace manually produced flowcharts even for programming purposes. They cannot, of course, ever replace systems flowcharts.

11
DECISION TABLES

This technique, which undeservedly has a high-sounding title, is both simple in concept and easy to use. It is by no means new, having been used in other fields—notably production engineering—for at least twenty years. Yet it was not until the late 1960s that it began to be used on even a modest scale in data processing. The impetus, when it came, was regrettably for the wrong reason. It became possible to convert decision tables directly into computer machine code, thus saving programming effort. This benefit, though real, is quite insignificant when compared with the true value of decision tables.

Decision tables are a very simple, yet effective, way of expressing the relationship between data, actions, people, programs, etc. As such, they can be of tremendous value to the systems analyst, who spends much of his time trying to do exactly that. Let us see how this is achieved by studying a very simple example. Suppose that, as a systems analyst working in the production information area, you establish the following facts:

(1) The production planning manager wishes to receive a report about any deviations of $\pm 10\%$ from plan in factory unit A.
(2) The sales manager wishes to receive a report about any falling behind of plan by more than 5% for product P.
(3) The standard cost sectional head must receive a copy of all production performance reports for product W in factory unit A.

Each of these statements is clear in itself. Yet in reading one after the other the relationships which obviously exist are difficult to grasp. A decision table showing the relationship is very simple to create and yet clearly shows the facts (see Fig. 11.1).

The general construction of a decision table can be seen from the above example and is illustrated in Fig. 11.2.

The steps in creating a decision table are:
(a) Identify the general conditions and list them in the left-hand upper part of the table.
(b) Identify the general actions and list them in the left-hand lower part of the table.
(c) Examine each required combination of general conditions, mark-them with a 'Y' (yes) or 'N' (no) when applicable, or a '—' (dash)

Condition/Action			
Deviation from plan ⩾ 10%?	Y	–	–
Factory unit A?	Y	–	Y
Product P?	–	Y	–
Deviation from plan ⩽ –5%?	–	Y	–
Product W?	–	–	Y
Send report to Production Planning Manager	X	–	–
Send report to Sales Manager	–	X	–
Send report to Standard Cost Sectional Head	–	–	X

Fig. 11.1. Simplified decision table.

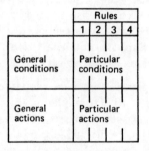

Fig. 11.2. General structure of a decision table.

if not applicable; for each set of conditions the corresponding actions are indicated by an 'X'.

A suitable *pro forma* for use in constructing Decision Tables is shown in Part II, 16.

However, the approach outlined so far is somewhat simplified. For instance, in rule 1 in the above example we considered the first two conditions :

- Deviation from plan ⩾ 10% ?
- Factory unit A?

which were sufficient to cover the case of the production planning manager. However, the combination of conditions :

- Deviation from plan ⩾ 10%?
- Factory unit A?
- Product W?

would also satisfy, if positive answers, the requirements of the standard cost sectional head. Clearly, therefore, the decision table becomes more elaborate if we start to show combinations of general conditions which satisfy combinations of general actions. Thus the example shown earlier becomes:

	\multicolumn{7}{c}{Rules}
	1	2	3	4	5	6	7
Deviation from plan ⩾ 10%	Y	Y	Y	N	Y	Y	N
Factory unit A?	Y	Y	N	Y	Y	Y	Y
Product P?	N	Y	Y	Y	N	Y	Y
Deviation from plan ⩽ −5%	−	Y	Y	Y	−	Y	Y
Product W?	N	N	−	N	Y	Y	Y
Send report to Production Planning Manager	X	X			X	X	
Send report to Sales Manager		X	X	X		X	X
Send report to Standard Cost Sectional Head					X	X	X

Fig. 11.3. Sample decision table.

This begins to indicate the power of a decision table. It can be used for more than just representing a set of logical relationships (highly valuable though this is)—it can be used as a *check on the consistency, accuracy and completeness of the analysis.*

For example, if we find certain combinations of general actions always occurring together, we have identified redundancy in the table. These associated general actions could therefore be combined, as logically they are one unit of action. We may sometimes find on our first draft of the decision table that certain general actions never seem to be needed. This is therefore a good pointer to re-analysis of the rules to see why a general action was earlier identified but later not used.

In this way the decision table is a general aid to the systems analyst in his analysis of the logical relationships in a system—always the most difficult part. Note, too, that these relationships can be at any level, for example, overall system logic, program relationships, data relationships, input–output-file relationships, etc.

One company known to the authors reserves decision tables for one special kind of use—interviews. After the systems analyst has gathered facts in his interviews with company staff, he summarises his findings in a decision table. He then revisits the persons interviewed, taking with him the draft decision table. He then discusses in detail the decision table itself with the company staff until final agreement is reached. A final version is then circulated as a statement of the system logic. The significant point here is the willingness and ability of a variety of company personnel to use and understand a formal method of logic analysis, in spite of the fact that they are in general completely unfamiliar with data processing. This brings out one of the great advantages of decision tables—the fact that they seem 'natural' to most people. Tables (e.g. bus time tables, price lists, etc.) are such a familiar form of expression in daily life—unlike flowcharting—that decision tables are readily accepted by almost everyone.

It should be pointed out, however, that a decision table contains *less* information than an equivalent flowchart. The basic concept of a decision table is simply the expression of relationships, and there is no significance normally attached to the sequence of conditions, actions and rules. Flowcharting also expresses relationships, but in addition presents a *procedure*, i.e. a sequence of actions and decisions is explicitly stated. This fact has given rise to problems in the conversion of decision tables to computer machine code. A good programmer will try to arrange his program so that it runs in the most efficient way possible. One way he does this is by drawing his flowchart. This saves the program following a long and wasteful path of decision points most of the time it runs. In creating a decision table, however, no attention is normally paid to the sequence of conditions, actions and rules. When converting the decision table to machine code, therefore, the conversion routine simply takes the rules in the sequence they appear, which is in general *not* the most efficient sequence.

Attempts to circumvent this problem have mainly been based on asking the analyst to provide probabilities of the outcome of the various rules and tests. This is an interesting approach, but it does complicate what is otherwise a simple concept. Most of the benefit of using decision tables can be obtained by using them in their simplest form. It is doubtful whether the pursuit of sophistication is worth the trouble for the average systems analyst. Nevertheless, the provision of the means to convert decision tables to machine code is a useful step towards reducing the overall workload of systems and programming.

There are two ways in which this conversion process to machine code is achieved: by preprocessor or by direct compilation. A preprocessor simply takes a decision table, punched in the appropriate format on punched cards or tape, and converts this data to statements in the computer language being used. For example, in the COBOL language a preprocessor converts decision table data mainly to 'IF' statements. These 'IF' statements are then included with the rest of the program statements as input to a normal compilation run using the usual COBOL compiler. A direct compiler, however, will accept the decision table data as input, so that no

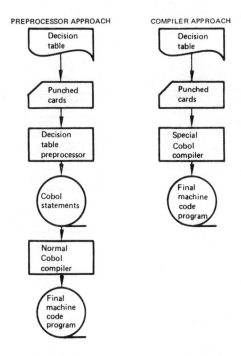

Fig. 11.4. Preprocessors and compilers for decision tables.

'IF' statements ever appear on the program printout, simply a decision table printout. The difference between the two approaches can best be illustrated by the flowcharts shown in Fig. 11.4.

Both these approaches are used and available through many of the computer manufacturers and software houses.

A practical problem which arises sometimes in the use of decision tables is that of size. It is inconvenient to stick pieces of paper together to create a large decision table, as may be necessary if there is a large number of conditions, activities and rules. A simple way round this is to split a decision table by making one or more of the conditions generate an action to go to another decision table as illustrated in Fig. 11.5.

In this way a structure of interrelated decision tables may be built (Fig. 11.6).

	Rules		
	1	2	3
Customer category A?	Y	N	Y
Cash < £100?	Y	–	–
Cash > £1000?	–	–	Y
Inform Sales Manager	X	–	–
Create new file			X
Inform Credit Manager			X
Go to decision table 'B'		X	X

Fig. 11.5. Subdividing decision tables.

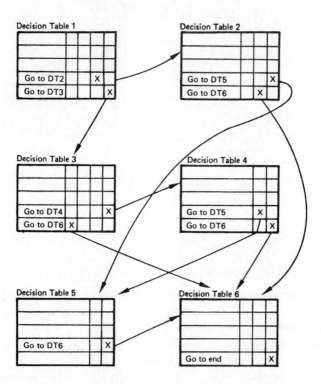

Fig. 11.6. Linking decision tables.

This is a 'flowchart' of decision tables. In a similar way decision tables can be embedded in a traditional flowchart. This is a very important point. Some people wrongly believe that decision tables can *replace* flowcharting. This is not so and never was intended to be the case. They are, in fact, best used in conjunction with flowcharting. This is because the merits of both techniques are complementary to each other. A comparison between the two techniques shows:

FLOWCHARTING	DECISION TABLES
• Good at showing procedural flows, less clear in showing logical relationships • Clumsy when introducing changes in logic • No check provided for logical consistency and completeness • Not easily understood by non-systems personnel	• Good at showing logical relationships; do not show procedural sequences • Easy to introduce changes in logic • Can be used to verify logical consistency and completeness • Shows logical relationships in a way easily understandable by non-systems personnel

It is therefore convenient to use flowcharts for the procedural parts of analysis and design, and decision tables to express the logical relationships. By intelligent use, decision tables can provide a valuable tool for the systems analyst in a large cross-section of his work, from the initial survey of overall requirements, through the system definition stage, to the actual programs produced in the implementation phase.

12
SIMULATION

Simulation is a tool which in the appropriate circumstances can be of immense value to the systems analyst. In general, simulation is used for the selection, analysis and design of medium to large systems. However, it can also be very valuable for certain aspects of even very small systems. Some of the ways in which simulation can be applied to a range of systems will be described in this chapter.

As its name suggests, simulation is the creation of an artificial situation which is meant to *represent* the real situation and then testing this artificially created one as though it were real. Clearly, the closer the artificial situation is to the real one the more useful the simulation could be. But it is usually expensive and difficult (and sometimes impossible) to use anything even approaching the real situation. So the skill in simulation lies in putting together a limited but reasonable representation of the real situation as simply as possible, so that it may be tested easily and modified easily but at the same time give meaningful results.

What do we mean by a 'real situation'? The example perhaps most familiar to readers will be the flight cabins of aircraft, which can be simulated by 'mock-ups' on the ground. Conditions in flight are simulated together with the effects of action by the pilot. This particular example brings out well the main benefit of simulation, namely *the ability to experiment without incurring the risks that would be incurred in the real situation.* There are many other physical simulators of this type, e.g. road traffic simulators. Simulation used in connection with business systems is essentially the same, even though the subject has been obscured with terms such as 'mathematical models' and the like. In principle, the process is exactly the same: a representation of the real situation is created, sometimes by means of equations, sometimes not, and then it is tested. For example, let us imagine we are working on a new stock control system and we have not yet determined which re-order rules best suit the company. We could construct a very much simplified version of the system, ignoring paper flows, amendment procedures, etc., and limiting it to one store only, just to try out different re-order rules to see the effect on the stock levels of that particular store. We could feed this simplified system, which does, however, represent the real system *in those aspects in which we are interested*, with input data representing the different situations we wish to test, e.g. periods

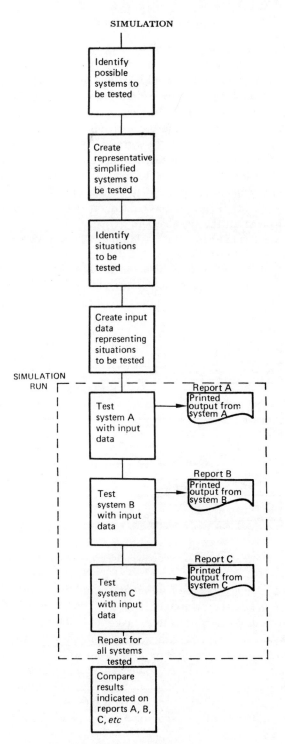

Fig. 12.1. Main steps of a simulation run.

of low demand, high demand, isolated large orders, etc. Our output from the simulated system would be the numbers of each product in the store at chosen intervals of time. This could show us how much money we have invested in stocks, how often we experience a stock-out, etc. In this way we can decide which of the re-order rules we tried is the best.

This last point brings out the big weakness of simulation. It can only give comparative results among the systems it is instructed to compare. It cannot give a guaranteed 'best' solution, i.e. the optimal solution. Simulation is merely a tool for testing different possibilities. If you overlook the best possibility, simulation will not point this out. The successful use of simulation depends therefore very much on the ability of the systems analyst to foresee all the possible alternatives.

How do we create a simulation run? The diagram shown in Fig. 12.1 indicates the steps to be taken.

System A, B, C, etc., may be exactly the same except for one or more decision rules which may be tested, as in the example of the stock re-order rule above. In this case the simulation run can be more compactly illustrated as follows:

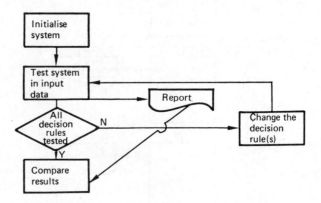

Fig. 12.2. Testing decision rules by simulation.

This idea of testing a range of different possibilities is, of course, fundamental to any kind of disciplined approach. What makes simulation a modern tool is its connection with the computer. This relationship is twofold. Firstly, the sheer speed with which a computer can work allows one to try out thousands or millions of possibilities which just could not have been humanly feasible. So the computer makes simulation of a large range of possibilities a practical proposition. Secondly, as computer systems are expensive, it is important to select the right one. So one tries to approach the selection task with all the analytical aids possible. Simulation is therefore used increasingly in the selection of computer systems by simulating one computer on another.

Because this last mentioned use of simulation is of particular importance, we will devote the rest of this chapter to various simulation techniques using the computer selection as an example. There are three commonly used techniques for this purpose:

12.1 Desk Simulation

This simply means a pencil and paper approach. The various alternatives are tried out by calculating the results for each possibility under the conditions selected. For example, in a financial analysis, different break-even points would be obtained according to different combinations of various financial facts, such as the interest rate selected. However, the most frequently used technique in the selection of computer systems is a points weighting system. Points weighting systems are based on the allocation of different values to the different factors involved in the selection, according to the relative importance attached to each factor. Let us imagine, for example, that we are choosing a computer system from five offers. We might be basing our decision on the six factors listed down on the left-hand column in the table below. To each of the factors we attach a relative importance value (starting by allocating '1' to the one of least importance). We are then in a position to give a points value to each system for each factor. We take each criterion (e.g. price, delivery date, etc.) in turn and rank each system according to the data we have about them. The highest value is attached to the system which has the most favourable characteristics, for example, lowest price. In our example below, system B is the cheapest, followed by systems A, C, D and E. So we attach ranking values to them as follows:

A (4), B (5), C (3), D (2), E (1)

for the price criterion. We then use these ranking values to multiply the weighting factor to give us our entries in the table as follows:

criterion	weighting factor	ranking					system				
		A	B	C	D	E	A	B	C	D	E
price	5	4	5	3	2	1	20	25	15	10	5
delivery date	2	5	1	2	4	3	10	2	4	8	6
maintenance support	3	1	5	2	3	4	3	15	6	9	12
systems support	4	5	4	3	2	1	20	16	12	8	4
effective core speed	1	1	2	5	3	4	1	2	5	3	4
software availability	3	3	5	4	2	1	9	15	12	6	3
						Totals	63	75	54	44	34

Fig. 12.3. Desk simulation using a weighting and ranking scheme
—I.

The system with the highest total of points (system B in our example) is then the best system in terms of the factors considered. However, we may not feel sure of our weighting scheme and so we may decide to try a different set of values, such as:

criterion	weighting factor	ranking					system				
		A	B	C	D	E	A	B	C	D	E
price	6	4	5	3	2	1	24	30	18	12	6
delivery date	1	5	1	2	4	3	5	1	2	4	3
maintenance support	2	1	5	2	3	4	2	10	4	6	8
systems support	4	5	4	3	2	1	20	16	12	8	4
CDU speed	3	1	2	5	3	4	3	6	15	9	12
software availability	5	3	5	4	2	1	15	25	20	10	5
						Totals	69	88	71	49	38

Fig. 12.4. Desk simulation using a weighting and ranking scheme
—II.

With this weighting scheme, system B is still the best system, though system C moves into second place instead of system A. We can try out, simulate, other weighting schemes to see how they affect our system selection.

The points weighting technique is very useful in a variety of applications, whenever one wants to rank and compare a number of similar items. It is not, of course, simulation in itself. It was chosen as an example for desk simulation, because it is so useful in the selection of computer systems, which are particularly relevant to the following two techniques, benchmarks and simulation by software. A points weighting system brings out very well, however, the principles and benefits of desk simulation, namely a systematic approach at comparing items which forces one to quantify factors whose assessment would otherwise remain as 'intuition'.

12.2 Benchmarks

These are a particularly interesting type of simulation, used increasingly in the selection of computer systems. The basic principle is simply to create representative tasks, which will be required to run on the selected system, and try them out on each of the considered systems. The performance of each considered system in performing the given tasks is compared in terms of time consumed, computer memory used, etc.

A primitive attempt at using benchmarks might simply be therefore to run the existing payroll program (if this is possible, i.e. if it is compatible at machine level through COBOL, for example) on each of the possible computer systems. This would be costly, because files, etc., would have to be set up, and not effective as an indicator to the best system, because the payroll program is almost certainly not representative of the overall data processing load.

This last point is crucial: a benchmark must by definition be representative of the workload it is simulating. The problem is therefore to decide on what is typical of the data processing workload, so that representative tasks can be created. The following steps outline the way in which benchmarks are created and used.

12.2.1 *Identify major computer applications.* Most computer installations have not more than three or four major applications which account for most of the productive time of the computer.

12.2.2 *Identify major computer programs.* The applications identified in point 1 above will probably contain anything from 10 to 200 computer programs. However, it is a well-known law (Pareto's Law) in the computer field that 70% of the computer time is consumed by only 15% of the programs. Be careful here not to exclude general-purpose programs, for example a generalised input program which may serve more than one application.

12.2.3 *Classify major programs.* The programs identified in point 2 will cover a variety of data processing tasks. It is now necessary to separate these programs into groups according to the type of processing they carry out. The major categories should be:

- input/output programs,
- calculation and process bound programs,
- file maintenance programs,
- utility programs such as sorts.

Individual companies will find they have certain types of programs in other areas, too, such as editing programs.

12.2.4 *Create representative tasks.* An analysis of each category identified in step 3 will enable characteristic tasks, or sets of computer instructions, to be created. For example, one representative task might be to locate a given record in a file and extract it. Another might be a series of instructions to check the validity of a piece of data.

12.2.5 *Build representative programs.* This step is simply the linking together of the representative tasks to create meaningful programs. It will almost always involve the concept of repetitive 'loops' within programs, that is the repetition of certain tasks many, many times.

12.2.6 *Create representative workloads.* To test the benchmark programs arrived at in step 5, it is clearly necessary to have suitable input data. This can only be created after the typical job mix is known for the installation, with various other local factors such as the dependence of one program on another. Peak loads should be created as well as the normal volumes.

12.2.7 *The simulation run.* Run benchmark programs on representative workloads. This step represents the culmination of the benchmark preparations. The computer system is tested by simulating representative programs with typical workloads. It is important to look for oddities here, as almost certainly mistakes will have been made in creating the programs. When the simulation appears satisfactory, the results can be taken and analysed.

The use of benchmarks requires considerable care and technical skill. The systems analyst will certainly not work alone and will need to work with an experienced computer programmer. Together they can use benchmarks very effectively to simulate their requirements on a new machine or a revised configuration and thus obtain an invaluable guide to the relative merits of proposed systems.

12.3 Simulation Entirely by Software

When using benchmarks, one builds representative tasks and tries them out *on the proposed machine* itself. It is sometimes not possible to do this, for example when a special configuration is being ordered (often the case with communication-based systems where particular combinations of telephone lines and terminals are used). Under these circumstances the computer system itself is simulated as part of the overall simulation procedure.

As indicated above, this is often necessary in the case of communication-oriented systems, especially as the workload is likely to vary greatly by time. Traffic on the lines (incoming and outgoing messages) will usually be random in nature and, although the overall pattern may be known, it is difficult to predict how any one proposed system will cope under the varying conditions. In this sort of situation, simulation is invaluable—and often the only way to assess system performance accurately.

To carry out such a simulation, one creates a model of the proposed system in software, that is a set of computer programs are prepared which will act as though the proposed system existed. This is not an easy task, but it has been made easier by the introduction of simulation 'languages', which are specially designed to describe computer systems, message flows, etc. The two best known languages are GPSS and SIMSCRIPT. Alternatively one of the specialised simulation packages such as SCERT or CASE may be used.

GPSS stands for General Purpose Systems Simulator and was developed by IBM for use on their machines. It consists of about thirty or so complex computer instructions or statements which are used to describe and define the system and the way it operates. It is especially suitable for 'queuing' situations, where messages arrive, wait and are then forwarded. It generates extensive reports of the performance of the system, such as:

- Number of messages handled.
- Average time to handle each message.
- Average number of messages waiting.
- Maximum number of messages waiting, etc.

SIMSCRIPT is a more general-purpose language, based on the mathematical programming language FORTRAN. It has more possibilities than GPSS because it is more basic, but for the same reason gives less help to the person creating a simulation program. It therefore requires more programming skill than GPSS. Both languages, however, have been widely used in a variety of applications and are especially useful in the simulation and selection of communication-based computer systems.

We will close this chapter by giving an example of a typical situation in which simulation entirely by software is used, probably utilising one of the languages mentioned above, GPSS or SIMSCRIPT. Let us assume that one is designing a simple data communication system with five terminals, which are distributed geographically as indicated in Fig. 12.5.

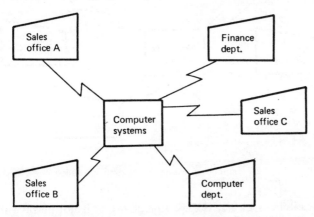

Fig. 12.5. Simulation example: ideal solution from user's viewpoint

The zig-zag lines indicate that ideally one wants to provide each place with its own direct link to the central system. However, this may not be necessary, because the Finance Department and Sales Office C use their link only infrequently and can probably share a telephone line without significant loss of service (that is, having to wait very often). So the systems analyst starts to consider the configuration shown in Fig. 12.6 as probably a more economic solution.

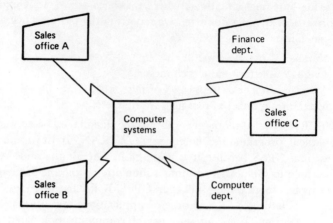

Fig. 12.6. Simulation example: alternative solution 1.

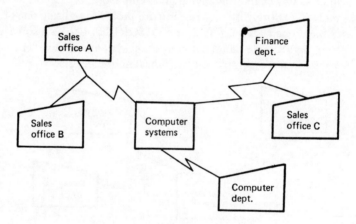

Fig. 12.7. Simulation example: alternative solution 2.

But the possibilities clearly do not stop there. Is it not possible that the configurations shown in Figs. 12.7 and 12.8 would also be acceptable?

The systems analyst wishes to select the configuration of least cost which will meet his requirements. To decide whether a given configuration is acceptable, therefore, he wishes to 'try it out'. This is where he can usefully turn to simulation. By specifying the various configurations above and applying various traffic loads to each in turn, the analyst can by simulation obtain a very good picture of the effectiveness of each system. He may find, for example, that one configuration gives longer waiting times for Sales Office B than another, but that it gives a better service to the Finance

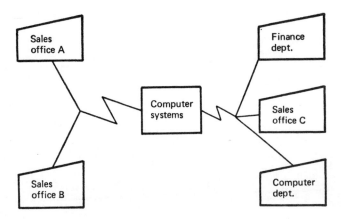

Fig. 12.8. Simulation example: alternative solution 3.

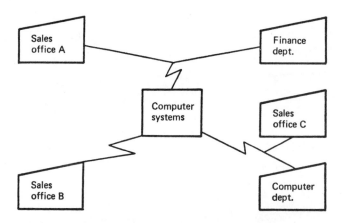

Fig. 12.9. Simulation example: possible optimum solution.

Department. With the information he obtains from the simulation run, he is in a good position to select the best system from those he has tried.

It is important to note, however, that the systems analyst may have completely overlooked the best possible solution, namely that shown in Fig. 12.9.

The simulation itself will not indicate that this is the optimum configuration, nor even that it is a possibility. Only if the analyst has specified that it should be tried will this configuration be selected. So let us end this section on simulation with a word of warning: simulation can be a very valuable aid for the systems analyst, but attention should be paid at all times to the fact that it never *guarantees* an optimal solution.

PART III
General Systems Considerations

13
DATA CAPTURE AND INPUT/OUTPUT

13.1 General Design Considerations

It is sometimes contended that data preparation is outside the scope of the systems analyst. There is certainly a widespread tendency for the data preparation aspects of a system to be given scant attention, because the analyst considers that it is more important to concentrate on the computer processing part of the job. Yet this attitude ignores the fact that more time can be wasted in a system by bad data capture techniques than can ever be saved by even the most elegant programming. Similar considerations can also be applied to input and output. Furthermore, lack of attention to the input phase of a system will almost certainly increase the number of errors entering the system. This latter point will be discussed in greater detail in the section on Data Security (Part III, 15).

What, then, are the objectives of the systems analyst with regard to data capture and input/output? They may conveniently be summarised as follows.

To reduce the total volume of input to the minimum practical level.

To reduce the amount of input that has to be manually prepared for any computer to the minimum practical level.

To design the input to the system in such a way as to ease the task of any person engaged in the preparation of that input.

To use the minimum number of steps practicable from the origin of the data to its input into the computer.

To reduce the volume of output (especially printed output) to the minimum consistent with achieving the aims of the system.

To design the final output of the system in such a way as to assist in the comprehension of the data.

To provide the output to the person who needs it.

We will now examine each of these points in turn.

13.1.1 *To reduce the total volume of input to the minimum practical level*

Clearly the less input there is the less preparation it will take. In addition, the percentage of errors is fairly constant in any process, so the lower the volume of input the fewer the number of errors introduced into the system.

How can the volume of input be reduced?

The first general rule is that only variable data should be entered, accompanied by the minimum of constant data. Thus, in a payroll system, the input giving the hours worked needs only to be accompanied by the paynumber of the employee working those hours, not the individual's name as well.

The second general rule is that repetitive data should be entered into the system only once. For example, in a payroll system, the week number (for tax purposes) and date (for identifying output) need only be entered once.

The third general rule is that only exception data need be entered into the system. An example of this can again be found in a payroll system. If the normal working week is 40 hours this information can be stored as a constant and hours worked only entered into the system for persons who depart from this norm.

The fourth general rule is to exclude spacing and editing operations from input which is to be key punched. For this form of input it is quicker if spacing is omitted completely. Editing, e.g. the insertion of points, can be implied from the format when the input is read into the computer.

13.1.2 *To reduce the amount of input that has to be manually prepared for any computer to the minimum practical level*

The key point here is that, where it is necessary to enter repetitive information into a system, measures should be taken to automate the process as far as possible. For example, it is desirable to have card codes in any punched card input merely to identify the card type (see page 97). Many installations reserve the first four columns of every card. The first two columns have a numeric code for the system and the third and fourth columns have a code denoting the card type within that system. It is not necessary for this information to be manually punched in every card, as it can be duplicated from a lead card during the keypunching operation.

13.1.3 *To design the input to the system in such a way as to ease the task of any person engaged in the preparation of that input*

There is more to this point than the simple desire to make the work as easy as possible for the people operating the system. The more complex the input the more errors in general it will produce and, furthermore, require more highly qualified (and therefore more expensive) staff to prepare it. A major aim of the systems analyst must therefore be the simplification of the clerical procedures in data capture. This point is further discussed under the heading 'Record Layout' in Part III, 14. The major points are the necessity for the person recording the data at source to do so in a logical sequence and for any transcription process, e.g. keypunching or keytaping, to follow the same sequence.

DATA CAPTURE AND INPUT/OUTPUT

13.1.4 *To use the minimum number of steps practicable from the origin of the data to its input into the system*

The ideal is, of course, the automatic capture of data at source in a computer intelligible format, e.g. punched paper tape. This ideal is not always possible or economic and alternative solutions will have to be adopted. In all cases a major design objective should be to minimize the number of stages through which the data must pass before input into the computer. The reasons for this objective are that every stage will occupy time, therefore increasing the total elapsed time between the creation of the data and the time that the final results are available to the users and, secondly, that each stage through which the data passes, whether it is merely transportation, transcription or preprocessing, significantly increases the possibility of errors in the system.

13.1.5 *To reduce the volume of output (especially printed output) to the minimum consistent with achieving the aims of the system*

During the past decade computer processing speeds have increased dramatically, but input and output devices have not kept pace with this development. This has meant an ever-widening gap between the speeds of computing and peripheral activities. This gap is a major problem in data processing. It is true that the advent of spooling techniques and the use of slave printers has improved the situation for the users of larger scale systems, but these are means of minimising the effects of the problem rather than solving it. For input the design criteria outlined above together with the introduction of faster data input devices (see page 90ff) offers some prospect of a significant reduction in the gap between input and processing speeds. At present the only devices available as alternatives to conventional line printers are considerably more expensive, as is the use of a slave computer for printing. In any case these alternatives are not viable for the users of small- or medium-scale machines which account for a large proportion of the total users. The only real relief of this problem at the output stage is *a reduction in the total volume of output*.

Furthermore, there is increasing evidence that the volume of output now being produced is beyond the ability of management to assimilate. The only way to increase the impact of the information supplied is therefore to *reduce* its total volume. One way in which the volume of data can be significantly reduced is by the adoption of exception reporting.

Exception reporting is the technique whereby only information which is abnormal is output from the system in human readable form. The remaining data is retained in machine readable form only. Thus, using exception reporting, a stores manager instead of receiving a complete report on the stock level of every item would only receive notification of those items where the stock levels were unusually high or low. This technique has two advantages: firstly, it reduces the overall volume of printed output and thus enables more efficient use to be made of computing equipment; secondly, it provides the user concerned with output which has been pre-analysed and those items which may require action are already identified without the need to search large volumes of printed output.

Although widely accepted in theory, exception reporting is by no means as widely used in practice. To accept such a system the user must have full confidence in both the system and in the data processing department and the analyst should seek to prove that such confidence is fully justified. A second way of achieving this objective is to make some reports available only on specific request.

13.1.6 *To design the final output of the system in such a way as to assist in the comprehension of the data*

In the final analysis the object of any system is to supply information to one or more users who may or may not be the persons who created the original data. A major, indeed perhaps *the* major, design consideration should therefore be the convenience of the user, the whole system being geared to meet his (or their) requirements. It follows that the output of the system should be designed in such a way as to facilitate its comprehensibility and utility to the user. This can be broken down into separate elements, i.e. the output should provide what the user wants, when the user wants it and in a manner in which the user can use it. Examples are legion of users patiently transcribing computer output manually into the form in which they require it. This is a complete waste of both time and money and is a condemnation of the systems analyst who designed the system or the data processing management who do not or cannot amend the system. This is not to say that the analyst should slavishly provide all that the user says he requires. Rather it demands that the analyst discovers what the *user really needs* and furthermore designs and documents the system in such a way that modifications to the output are easily made.

13.1.7 *To provide the output to the person who needs it*

This simple point is often neglected, but by neglecting it a severe penalty may be paid in the time needed for action to be taken on the result of a system. Output may be either operational or advisory. Operational output is that which causes action, e.g. a report on significant variations from budget should initiate investigation of the causes. Advisory information keeps people 'in the picture', e.g. the staff welfare department being advised of the names of all new staff.

The requirements for detail contained in output will vary between these two classes and generally with the status of the recipient. In general more senior staff have less requirement for detailed information and more requirement for summarised information.

13.2 Input Methods

The term input as used in this chapter covers all the stages from the recording of the basic data to the feeding of this data into the computer for processing. The term data capture is also sometimes used for this purpose. The basic steps in this process are :

- Recording.
- Transport and collection.
- Preparation, e.g. keypunching.
- Verification.
- Sorting.
- Control.
- Computer input.

In any particular application these functions may not all exist or they may take place in a different sequence from that shown. As indicated in the section on general design considerations above, major design considerations should be to reduce the volume of the input and to reduce and automate the stages in its introduction into the computer system.

In any particular system the choice of input procedure adopted will depend on the following considerations :

- The volume of data to be expected.
- The time limits imposed on the system.
- The minimum level of accuracy demanded of the input.
- The cost.

The major methods used for data input are :

13.2.1 *Punched cards*

The conventional data processing punched card is designed to contain up to 80 characters of information which may be numeric, alphabetic or one of a range of special characters. Other formats in use include cards containing 40, 56, and 96 columns of information. These cards can be punched by a variety of devices, but the most common are hand or electrically driven keypunch devices and mark sensing equipment (see Fig. 13.1).

Keypunches are simple mechanical or more commonly electromechanical devices which enable the operator using a keyboard similar to that of a typewriter to punch the appropriately coded rectangular holes into a card. Some types store the information until the complete record is assembled and then punches all the data while other types punch column by column.

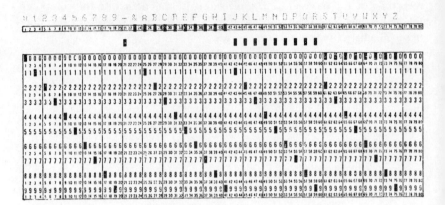

Fig. 13.1. An 80 column punch card showing the representation of numeric and alphabetic characters.

Verifying is the technique whereby a second operator performs essentially the same actions as the person who originally punched the data using the same source documents and the previously punched cards. When the key depressed by the second operator does not agree with the key depressed for the same characters by the first operator the verification machine 'locks' or otherwise signals the difference. The second operator must then examine the difference and either check the source document or (more usually) repunch the card. This check does not, of course, prevent two operators misreading the same source document but does guard against the majority of accidental errors by operators. After successful verification the machine normally places a small notch in the edge of the card to denote that verification has been carried out.

With mark sensing equipment the information to be punched into the card is marked manually on to the face of the card using a pencil or ink with a high carbon content. A card face showing the spaces into which the pencil or ink marks are made is shown in Fig. 13.2.

The card is then inserted into a special card punch which 'reads' the marks by passing electrical currents, so that the presence or absence of the high carbon content mark completes or fails to complete the circuit, thus determining whether or not a character is punched in any particular

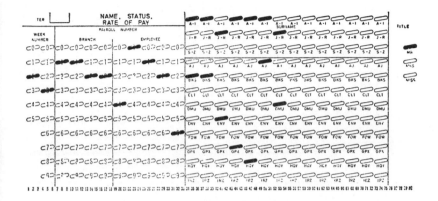

Fig. 13.2. A mark sense card showing both the original handwritten and machine readable data representations.

column. This provides a very suitable way for obtaining data at source, because of the simplicity of the completion process and the convenience of the punched card as a unit medium. Unfortunately this process is very susceptible to error caused by smudging the marks on the card and other superficially minor damage to the card. The number of successful applications is therefore limited and the systems analyst is strongly advised to hold extensive trials *under actual working conditions* before introducing a system based on mark sense reading.

The general advantages of punched cards as an input medium are:

(1) The ease with which faulty cards may be replaced and the ease with which card files may be updated by inserting or removing cards.

(2) Punched cards are a form of computer input which is directly intelligible to man and machine. In addition to punching the appropriate codes into the card, many keypunch devices also print the character on the top edge of the card. If this facility is not incorporated in the card punches in use, separate interpreters can be used for the same purpose.

(3) Punched cards may be used as working documents. Because punched cards are by nature unit documents, they are particularly suitable for use as working documents, e.g. when an invoice is sent

to a customer it may be accompanied by a punched card which the user is requested to return with his cheque. The returned card can then be used as a notification of payment received after only a quick check that the accompanying cheque corresponds in value to the invoiced amount. Mark sense systems (see above) are also examples of the use of punched cards for working documents. To facilitate the use of punched cards in this way a wide variety of card types, including cards with stubs attached and cards made into sets with forms and carbon paper, are available from the suppliers and they can be printed to suit individual requirements.

(4) By virtue of their unit document nature, punched cards are particularly suited to off-line processing on electro-mechanical equipment. Among the tasks which can readily be performed are sorting, counting and merging. Although this facility is unlikely to be a major consideration for the user of a large-scale computer, it may well prove to be extremely important for users of small- to medium-size computers.

The general disadvantages of punched cards are:

(1) The relatively slow process of punching and verifying. It is normally stated that a keypunch operator who is suited and accustomed to her work will complete approximately 8,000 keystrokes per hour. This figure will, of course, vary according to the nature of the punching (numeric, alphabetic, etc.), the legibility of the source documents and the number of columns punched in each card. However, using the above figures as a guide, it may be stated that 400 cards fully punched and verified would be a day's work for a punch operator. This would take a medium-speed card reader approximately 1 minute to read into a computer!

(2) Although in the example quoted above the speed of the card reader is very fast compared to the speed of the human operator, the speed of the card reader is itself very slow compared to the speed of a computer. A card reader rated at 400 cards per minute has a maximum data transfer rate of only 533 characters per second.

(3) The number of characters which can be contained in one card is low. By the nature of the medium the number of characters which can be contained in the card is limited to 80 or whatever limit is set by the type of card in use. This may mean that more than one card is needed to contain certain records with the consequent inconvenience of having to repeat identifying information.

(4) The very ease with which a card file or deck can be amended does, of course, mean that it is easy for an individual card to become separated from the deck to which it belongs.

13.2.2 *Punched paper tape*

As its name implies, this medium consists of a strip of paper tape usually about 1 inch (2.54 cm) wide which is punched in a manner analogous to punched cards (although circular holes are normally used). A smaller row of holes along the length of the tape and slightly to one side of the centre line is used for guiding the transport of the tape. Codes using 5, 6, 7 or 8 holes (or spaces) punched across the width of the tape are in common use. One of these rows of holes is usually reserved for a parity check (see Part III, 15).

The advantages of paper tape are :

(1) Because of the nature of the medium, the length of each record is effectively unlimited.
(2) Computer input can be faster than that obtainable with punched cards. This stems not only from the physical properties of the peripheral devices, but also in the amount of medium which is not recording data which has to be passed through the reader. With a punched card the entire 80 columns must pass under the reading mechanism even if only one column contains a character. Thus much of the theoretical capacity of a card reader is usually lost reading blank columns. With paper tape, on the other hand, only a short inter record space or end of record marker is needed between records so that the minimum of time is wasted 'reading' gaps between data.
(3) Paper tape is easier to store than punched cards because of its more compact format.
(4) Paper tape is well suited to data transmission by telex.

The disadvantages of punched paper tape are :

(1) The slow preparation of paper tape.
(2) The relatively slow input to the computer. Although it is practically faster than punched card input, paper tape input is still very slow when compared to the speed of a computer.
(3) The difficulty of making amendments to paper tape files. This is the most significant drawback to the use of punched paper tape. Insertions have to be performed by copying the original up to the point at which the insertion is needed and then punching in the additional information. Verification is also performed by a copying process in that the original tape is read and the character com-

pared with that keyed by the verifier operator. If they agree a new tape is punched; if they do not agree the verifier locks. Deletions can be made more easily by punching over the existing characters to produce a character which is recognised by the tape reader or program as not for processing (all the channels are punched).

Both paper tape and punched cards have their advocates, particularly in regard to their relative robustness. In addition to direct preparation it is possible for both media to be created as a by-product of some other process performed by a business machine, e.g. many accounting machines produce a paper tape record of some (or all) of the data entered on the keyboard or calculated by the machine. This paper tape can (provided compatible codes and formats are used) be used directly as input to a computer program.

Automatic converters are also available to enable certain types of encoded data, e.g. identity badges or the Kimball Tags often used as identity tags in clothing offered for sale, to be translated into conventional punched cards or paper tape. Some of these converters allow two or more encoded data sources together with some standard data, e.g. the time and date to be punched into one card or paper tape.

The advantages of systems of this type are that the total time between origin of data and its availability for processing is reduced and, secondly, the data input into the system is as accurate as the source information.

13.2.3 *Keytaping*

A number of devices are now available which enable an operator to enter data directly on to magnetic tape. These devices may be divided into two categories: firstly, those machines which work independently so that each operator is working with a separate reel of tape and, secondly, those systems where a number of operators are all working through a control device (effectively a small computer) which may or may not format and check the data to enter data on to one tape.

The advantages of keytaping as a means of preparing input are:

(1) It is faster than punching either cards or paper tape.
(2) There is no difficulty in accommodating variable length or very long records.
(3) The resulting magnetic tape is capable of being read into the computer at much higher speeds than either punched cards or paper tape.
(4) Noise level. As these machines are not electro-mechanical, they create much less noise.

There are, of course, also practical disadvantages. These may be summarised as:

(1) Difficulty in correcting errors. In many cases the operator will be aware of a mistake that has been made. The correction of this type of error is more difficult with a medium on which the operator cannot see what he or she has keyed in. The reading devices currently available with keytaping machines are not as easily understandable as punched holes in tape or card.

(2) Utilisation of tape. The daily output of a single keytape operator can be accommodated easily on a few feet of magnetic tape. Since a normal reel of tape is hundreds or even thousands of feet long, this leads to very inefficient use of tape storage devices. This is a problem in its own right, but a more serious problem is in batching work through data preparation. Where turnaround times are vital, it will probably be necessary for many data preparation staff to be working on the same problem at the same time. Suppose for example that ten staff each spend a particular morning preparing input tapes each of which contains only a few feet of data and a consequent problem for the operations staff who will have abnormal demands for loading and unloading tape drives. Such considerations may well affect decisions about the peripherals needed in the computer configuration. This problem can be partially overcome by using a system whereby many operators are linked through a common device so that they put data on to one tape, or by use of a system which uses small non-standard cassette tapes which are then fed into a special on-line tape reader.

13.2.4 *Key to disk*

The factors for this type of input preparation are the same as those given above for keytaping. The data transfer rate for disks is greater than that for tape, so a further benefit accrues to the key-to-disk procedure in this respect. Because of the greater expense of disk units (both the removable pack and the drive unit itself), these devices usually work through a control unit which has data formatting and checking capability.

13.2.5 *Optical character recognition (O.C.R.)*

This system relies on optical processes to 'read' data which is in a form readable to human beings and to translate this directly into computer readable form. Such devices may be either on-line or off-line in which case some intermediate data storage is used, e.g. magnetic tape. Limitations exist on the range of characters which can be read by all the equipment currently available and most require highly stylised type faces. Those machines which read handwriting also are limited in this way and there are no machines currently available which read normal handwriting.

The principal advantages of O.C.R. systems are :

(1) The eliminations of the transcription and verification stages of the total input process. In some cases, however, it has been found necessary to employ a transcription process to type the source documents in order to produce suitable machine readable documents.
(2) Conventional sources of data can be used for computer input, e.g. adding machine tally slips.
(3) It is cheaper than either keypunching or keytaping. Assuming no transcription process is involved, significant savings can be achieved through the use of O.C.R.

The major disadvantages of O.C.R. systems are :

(1) O.C.R. readers demand very high quality input. Current equipment is very critical with regard to paper reflectivity, character formation, printing registration, ink used, etc. There is, therefore, likely to be a high reject rate in systems using O.C.R. techniques, especially if any attempt is made to read handwriting.
(2) The use of stylised type fonts makes O.C.R. documents less acceptable to users.
(3) Limited character ranges.

13.2.6 *Magnetic Ink Character Recognition (M.I.C.R.)*

In general characteristics M.I.C.R. equipment is analogous to O.C.R. equipment, but uses magnetic as opposed to optical techniques for reading the data. The information to be read in this way is printed in special magnetisable ink using a special type font. Reading is a two-stage process. The first stage is to magnetise the ink. The second stage is to produce an electrical current from this magnetised ink and the characteristics of this current enable the character to be identified. The most widely known application of M.I.C.R. is for the identification of cheques.

The advantages of M.I.C.R. are :

(1) Reading equipment is faster and tends to be more reliable than O.C.R. equipment.
(2) The elimination of transcription and verification of data.

The major disadvantages of M.I.C.R. systems are :

(1) The use of stylised type fonts.
(2) The high accuracy of printing required, i.e. size, separation, ink covering and clarity of characters.
(3) Some type fonts only cater for numeric information.

13.2.7 *Terminal devices*

In addition to the devices listed above a number of terminal devices are now available which can be used for data input purposes. These can be conveniently divided into keyboard and special purpose devices, the latter category including light pens, 'trackballs', 'joysticks', etc. In general it may be stated that these devices are not suited to the entry to the computer of the large data volumes associated with most routine data processing applications, being both too slow (i.e. limited to operator speed) and as yet too expensive for such use, but this situation is expected to change.

13.3 Output

The term 'output' covers all the stages between the production of the results in computer-understandable form to their presentation to the user in a form that he can understand. The basic steps in this process are:

- Output in human-understandable form.
- Ancilliary operations (de-collating, bursting, trimming, etc.).
- Transport.

The basic methods used for producing computer output in human-understandable form are:

13.3.1 *Line printers*

This is by far the most important category of computer output devices. Characteristics of machines vary between manufacturers, but in general they work at speeds up to about 2,000 lines per minute, each line consisting of up to 160 characters. All the characters on one line are printed simultaneously. These devices may be either on- or off-line. Off-line printers may be single application machines or relatively slow (and therefore cheaper) general-purpose computers used for printing output (known as 'slave' computers). Slave computers and some off-line printers can edit and format the output before printing. Spooling (simultaneous peripheral operations on-line) is the technique whereby a computer will write output on to magnetic tape or disk and subsequently print it while processing another job. In this way the use of peripheral devices (including the printer(s)) is more nearly optimised. Normal line printers are capable of producing up to six copies of a document at one time either by the use of interleaved one-time carbon paper or non-carbon-required paper. Suitable computer printer stationery can be printed and perforated to the specification of the user and small documents (i.e. those which are less than the full available print width) can be printed two or more at a time.

13.3.2 *Character printer*

Unlike line printers, character printers print one character at a time across the page. Except for some small or mini-computers which are not designed to handle large data volumes, character printers are not used for general output purposes. Their use is usually confined to the printing of systems or log information. Their future use for the printing of *ad hoc* information from a data base is likely to grow as they are often part of a simple terminal device.

13.3.3 *Visual display devices*

Here the information appears on a screen very similar to that on a television set. The use of optical output devices is not yet widespread for commercial data processing systems mainly because of the relatively high cost. Like character printers, however, their use is spreading fast with the increasing use of information systems and large-scale data bases, as management increasingly require rapid access to this data. The devices are convenient to use and are dropping rapidly in price so that they are becoming increasingly attractive.

13.3.4 *Off-line peripheral devices which produce microfilm directly from magnetic tape*

These devices are well suited to the information storage and retrieval type of application, but are rather less relevant to the normal commercial data processing systems. The big advantage of such devices is the very high speed at which data can be turned into output. The disadvantage is that a special reading device is necessary before the output can be used.

13.3.5 *Punched Cards*

Punched cards can be produced as output either for use as control documents for numerically controlled machine tools or as a link between computer systems, e.g. a prepunched card giving details of part number, description, etc., may be produced as output from a production planning system. This could be used as a job ticket for the production of these parts and after the addition of further data, e.g. actual number produced, could be used as input to a job costing or stock system.

13.3.6 *Punched Paper Tape*

This is sometimes used as output but normally only for direct input to another computer system. Being a continuous rather than unit medium means that it does not have the versatility of punched cards for output use. Numerical control machine tools are one application of punched tape output.

13.3.7 *Plotters*

On-line or off-line plotters are available which will enable graphs, plans and isometric drawings to be produced by computer. Their main application is in the scientific and engineering fields.

13.3.8 *Magnetic Tapes*

These are often produced as output but normally only for use in an off-line output device, e.g. an off-line printer or for re-input to the same system at a later date or as input to another system, which may or may not be run on the same installation.

13.3.9 *Digital-Analogue-Converters*

A digital computer can produce output messages which can be used via an analogue converter to control a technical or scientific process.

14
FILE ORGANISATION AND RECORD LAYOUT

The layout of a record is the physical arrangement of the data elements which constitute that record. In designing the record layout, the analyst must cater for a number of differing requirements. Firstly, he must cater for the person or machine which creates the source data. Secondly, he must cater for the requirements of the process that translates that source data into machine-readable form. Thirdly, he must cater for the requirements of the storage medium being used. Lastly, he must cater for the processing requirements for which the data will be used. These requirements will often be different and may present unresolvable conflicts. In this case the analyst is well advised to cater for the requirements in the order in which they are listed above, since it is in the last resort the acceptance by the user of the system and his ability to use the results (largely conditioned by the ease of entry of data to the system and the accuracy of the results) which will determine the success or failure of the system. In this respect, therefore, the analyst must identify himself clearly with the user of the system in preference to the convenience of the programmer or the part of the system internal to the data processing department.

For the person creating the source data, the natural order for it to be recorded will normally be the chronological order in which it is created or arrives, or is extracted from other sources, or some natural association. Thus in recording personnel data, name, address, date of birth would be a 'natural' order, but date of birth, address, name would not.

Capturing data in a natural sequence is a significant step in reducing input errors as the requirements of the system are reinforced by the conditioned pattern of thought. Similarly the form or other medium on to which the data is entered should be arranged in natural sequence, i.e. left to right across the page (or top to bottom). When data is translated into machine-readable form (e.g. by keypunching or keytaping) the volume of errors can be significantly reduced by letting the operator follow the existing order of the data. The volume of errors will rise considerably if the punching instructions are for data elements located in random sequence on the source document. These factors will naturally tend to create a tape or card record which is in the same order as the original data was recorded and this will also be true for those data capture techniques which eliminate the manual step of translation into machine-readable form (e.g. Optical Character Recognition and the production of punched paper tape as a by-product of another process).

Having established that the layout of the records is of subsidiary importance to the requirements of the user of the system, the analyst should observe certain principles in the design of records. For the sake of convenience, these are classified by the storage media used.

(1) *Card Records*

The normal data processing punched card (Fig. 13.1) can contain up to 80 columns of information—each column representing one alphabetic, numeric or special character. The conventional coding to represent the alphabetic and numeric characters is also shown in Fig. 13.1. Other punched card formats contain 40, 56 or 96 columns of information (see Part III, 13).

Punched cards have two distinct uses for file purposes. Firstly, they may be used as the file storage medium and, secondly, they may be used as the means of creating a file on another medium, e.g. a magnetic disk.

For either use the major considerations to be remembered in designing card record layouts are :

- Each card must contain both a file and data identity. Consisting as it does of a number of separate individual cards, a card file must incorporate sufficient identity on every card to ensure that it can be returned to the correct position if it is removed or displaced from the file. Thus it must contain an identifier to indicate to which file it belongs and also an identifier to locate its position within that file.

  The file identifier (often called a card type code) should be the subject of an installation standard (see Part III, 16) and located in a fixed position of the card (e.g. the first four columns) and allocated by a particular person, e.g. the person responsible for standards or the chief programmer, to prevent duplication. Colour coding of cards may prove useful to prevent the accidental mixing of card files but does not eliminate the need for a file identifying code.

  The identifier to locate the position of an individual card within the file will usually be part of the data, e.g. personnel number or part number.

- Record length. There is a natural tendency with punched cards to try to limit the size of the record to the capacity of the card available, i.e. 40, 56, 80 or 96 columns. Where this is not possible, each card must contain the identifying codes above (this will entail repeating both the file and record identifiers) and to facilitate off-line sorting, these should occupy the same positions as on the first card. The card should also contain a code to indicate its sequence within the record, i.e. if a record contains 190 characters and is contained on three 80-column punched cards, these three cards for each record should be coded (say) 1, 2 and 3 to ensure that the record can be maintained correctly. When a variable number of punched cards are used for each record it may simplify processing if a last card of record code is used.

A suitable form for use in designing card record layouts is shown in Fig. 16.5.

(2) *Tape Records*

Tape records, either magnetic or paper, are by nature of the medium contiguous. It is not therefore necessary to include on every record information which is common to every record in the file (e.g. the file identification code). This information can be stored in a separate file header record which may also include information about the retention date associated with the data in the file.

Magnetic tape also provides the possibility of coding data in a more machine-oriented manner than is possible using conventional punched card techniques. Full details of these techniques will be found in the manufacturers' literature.

Furthermore, both kinds of tape media offer the possibility of virtually unlimited record length. The analyst, however, should ensure that the record length adopted is such as to enable efficient transfer of data by the effective use of the buffer areas available (if applicable to the computer being used) and are not unwieldly for processing—this latter point will be of particular importance on machines with very limited core.

A suitable form for use in designing tape record layouts is shown in Fig. 16.7.

(3) *Direct Access Devices*

The factors quoted above in connection with magnetic tape records are also applicable to other types of magnetic storage media.

The consideration of record size may, however, assume greater importance in relation to the efficiency of transfer of data and storage usage (see also page 110).

Consider a disk unit in which a sector of 320 words is read at a time. For a two-surface disk with eight sectors, this means data can be transferred at a theoretical rate of 8 times 320 per rotation of the disk, i.e. 2,560 words per rotation. If fixed-length records of 160 words are used (or any number of words which will divide into 320 without a remainder), this is also the effective data transfer rate. If, however, fixed-length records of 161 words are used and (as is true with some disk units), it is not possible to have a record located in two separate sectors, the effective data transfer rate will be only 8 times 161, i.e. 1,288 words per rotation, i.e. only 50.3% of the theoretical rate. This percentage is also the storage efficiency.

14.1 File Organisation Structures

There are three basic types of file organisation structure, namely:

Sequential.
Random.
List.

All other structures are derived from these three basic structures or from a combination of them.

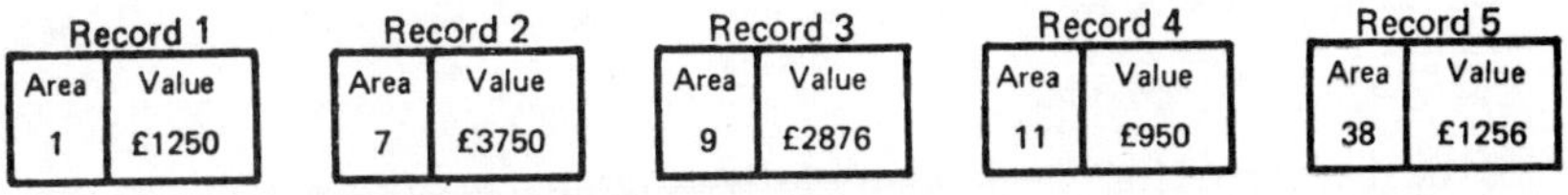

Record 1		Record 2		Record 3		Record 4		Record 5	
Area	Value	Area	Value	Area	Value	Area	Value	Area	Value
1	£1250	7	£3750	9	£2876	11	£950	38	£1256

Fig. 14.1. Sales data file organised sequentially by area code field.

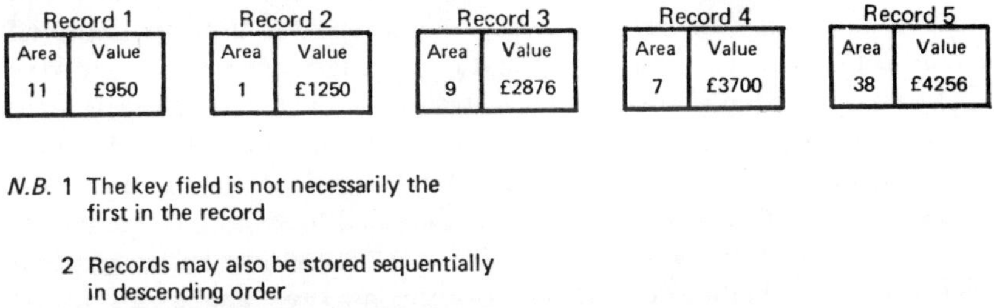

Record 1		Record 2		Record 3		Record 4		Record 5	
Area	Value	Area	Value	Area	Value	Area	Value	Area	Value
11	£950	1	£1250	9	£2876	7	£3700	38	£4256

N.B. 1 The key field is not necessarily the first in the record

2 Records may also be stored sequentially in descending order

Fig. 14.2. Sales data file organised sequentially by value of sales field.

14.1.1 Sequential Organisation

This is the conventional method of organising files by arranging them in sequence on the basis of a common attribute. The field containing this attribute is called the key field or simply the key of the record. Thus a simple file of sales data containing only area code and value of sales could be organised sequentially on either field. (Figs. 14.1 and 14.2.)

This form of organisation is, of course, the form adopted for the majority of clerical record-keeping activities and it was therefore natural that it should have been preserved when these applications were transferred to punched card machinery. The advent of magnetic tape devices which read the contents of a tape sequentially further perpetrated this philosophy. The advent of direct access storage devices, however (magnetic disks, drums, data cells, etc.), gave the possibility of directly accessing data stored on any part of the device without the necessity to read the complete file sequentially. To do this, however, required a fundamentally different approach to the concepts of file organisation.

One way in which file designers sought to use this direct access capability was the *binary search*. This technique, which is used with sequential files stored on direct access storage devices, is illustrated in Fig. 14.3.

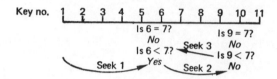

Fig. 14.3. Search to find key no. 7 using the binary search technique.

Basically this technique consists of dividing the file into two by selecting the mid-point and ascertaining in which half the required record is located. That half of the file is then bisected and the quarter of the file, in which the record is, ascertained. This procedure is repeated until the required record is located. It can be proved that the average number of record keys that have to be read is less than when straight sequential searching is used.

The Indexed Sequential System does to some extent embody the idea of the binary search technique in the formal organisation structure of the file, since a set (or sets) of indices are maintained to avoid the necessity of a straight sequential search. The resulting structure may be likened to a dictionary. The highest level index gives the last key entry in each of the next lowest level physical subdivisions of the file. At each level the index is searched sequentially until the first key value above the required key value is found. The associated hardware address directs the search to the relevant portion of the next lower level of index or to a portion of the file itself. This procedure is shown diagrammatically in Fig. 14.4.

The level of indices normally correspond to the physical organisation of the storage device, e.g. for a magnetic disk separate indices are constructed for each track and a higher level index constructed for each cylinder.

Provision has to be made for insertions into the file and this is usually by means of an overflow area and the creation of an overflow entry key associated with each entry in the lowest level index.

Care must be taken in the physical organisation of the file to prevent

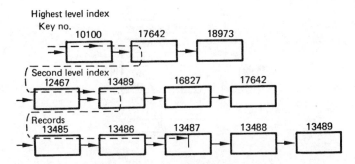

Fig. 14.4. An indexed sequential file structure. The search to find record 13487 will follow the dotted line.

excessive movement of the read/write heads of the device being used. Ideally each level of index above the level of track index should be stored on separate devices. In this case, if the file is accommodated on a single cylinder, all arm movement is eliminated during the processing of the file.

The major problem with the sequential and indexed sequential structures is that the file is, of necessity, arranged in sequence on a single key, e.g. personnel number, part number, etc. If the file is required in any other order, either a lengthy sort is required or alternatively a file of identical data organised in a different sequence must be kept. Furthermore, if only a small percentage of the records are used during normal processing, much time will be wasted reading unwanted records.

14.1.2 *Random Organisation*

A more fundamental realisation of the opportunities of the direct access devices led to the development of random file organisation. In this case, it is, strictly speaking, incorrect to speak of a structure since the file is dispersed throughout the storage device in a truly random manner. Although this dispersion is random, there must, of course, be some manner for the data to be retrieved if the system is to be of any practical use. There are three methods adopted for maintaining this control.

The first method is for the programmer to retain all the information about where each individual record has been located. There can, however, be little practical virtue in this system and its use cannot be recommended.

The second method relies on the use of a dictionary. Using this system (Fig. 14.5) a sequential dictionary is maintained showing the key and location for each record.

It is, of course, possible to have several dictionaries for the same file, thus overcoming one of the principal objections to sequential files. The alloca-

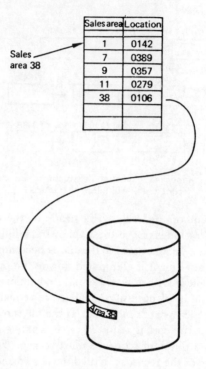

Fig. 14.5. Dictionary look-up of a random file.

tion of physical addresses to records when using this system can be by any convenient means.

If the dictionary method is used, the dictionary should be retained on the fastest possible device (or in core) to avoid losing the fast access time possible using random organisation.

The third method of controlling a random file so that retrieval is possible is to use a formula to convert the key into a hardware address (Fig. 14.6). This system, which obviates the need to maintain dictionaries, does, however, require special care as it is unlikely the formula chosen will always produce unique results.

This problem can be partially overcome by more elaborate formulae or by the use of a suitable overflow technique. A further danger is in the creation of a unique address which overwrites part of an existing record. Whenever a record needs to be retrieved, the same calculation as was originally used to create a storage address is performed and the result is the location of that record.

FILE ORGANISATION AND RECORD LAYOUT   103

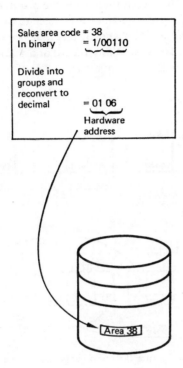

Fig. 14.6. A calculation method for determining the location of a record in a random file.

14.1.3 *List Structures*

The third of the basic organisation structures is the list structure. In this structure the logical and physical organisation of the file are divorced by the use of pointers. Thus each physical record includes a pointer to the next logical record. The file can, therefore, be created by adding each new record sequentially to the end of the file (i.e. the file becomes effectively a sequential file in which the (implied) key is the date of addition to the file). The logical sequence of the file is maintained by amending the pointers in the affected records in the file and incorporating a pointer into the new records. (Fig. 14.7.)

It is, of course, necessary for the file header record to incorporate a pointer to the first record.

More than one pointer can be incorporated in each record and this gives the opportunity to create multiple lists through one file of data. (Fig. 14.8.) To the user this has the effect of giving him a number of files incorporating the same data created in different sequences with consequent economy in sort times and storage requirements and the elimination of the problem of ensuring that every file is consistent with every other file. A list structure in which the last pointer directs the search back to the start point is known as a *ring structure* (Fig. 14.9) and means that, instead of referring to a dictionary, the user may look at any record and follow the pointer associated with a particular key field and this will lead to the required record.

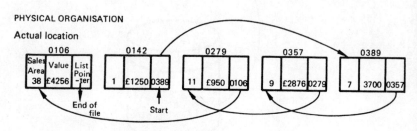

Fig. 14.7. List structure (logical organisation by sales area code).

Lists may be created to point 'backwards' through the data as well as 'forwards' and both may exist for the same data element.

Another extension of the list structure is developed by removing certain fields from the individual records and leaving only the pointer. The resulting structure which is called a *partially inverted list structure* is shown at Fig. 14.10.

Provided that multiple entries of the removed field exist in the file and that the pointer is smaller than the data element, this reduces the storage space required.

If this principle is carried out fully and only one data element remains in the record, the file organisation is described as an *inverted list structure*. (Fig. 14.11.)

The major advantages and disadvantages of each type of file structure are tabulated in Fig. 14.12.

14.2 Facts Governing the Choice of File Organisation

The factors governing the choice of file organisation system for any given application are complex. There are many factors affecting the choice and

FILE ORGANISATION AND RECORD LAYOUT

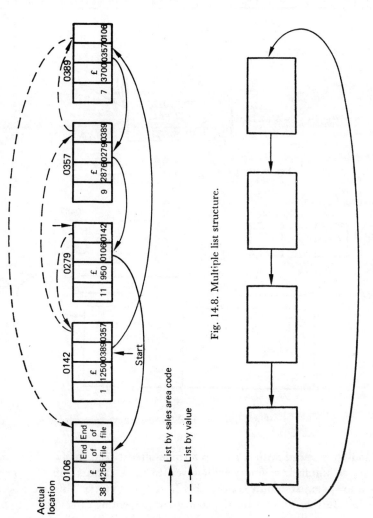

Fig. 14.8. Multiple list structure.

Fig. 14.9. The ring structure.

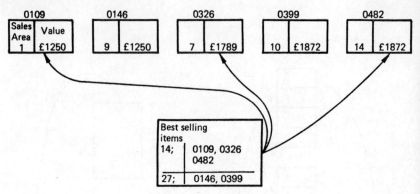

Fig. 14.10. Partially inverted list structure.

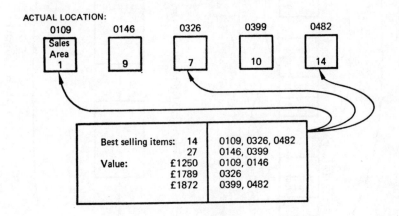

Fig. 14.11. Inverted list structure.

the analyst is faced with a multiple trade-off between many, often conflicting, requirements. The weight given to these factors will vary from system to system and quite conceivably from time to time. This latter point is, in fact, further support for the concept of preventing direct access to the files by the application programs by the use of independent data base management software. The major factors concerned in this multiple trade-off situation, which must be evaluated by the analyst, *in every case* are :

Storage media availability.
Hardware.
Programming possibility, ease and efficiency.

Operating system.
File size.
Record size.
Method of interrogation of the file.
Updating frequency and volume.
Response time requirements.
Cost.
Space management considerations.
File security considerations.

14.2.1 *Storage Media Availability*

Obviously the first factor the systems analyst must consider is which medium is available on the machine on which the system is to be implemented. This goes beyond the simple question of: 'Are there tape decks or disks on this machine?' The analyst must know which of the peripherals are reserved for the operating system, spooling, etc., before he can make a decision on this point. He will also need to decide the means of updating his files—whether by copying or overlaying since this will affect the number of devices that will have to be used.

It may be, of course, that, even if the peripherals required are not currently available on the computer, a strong case can be made for acquiring them. The analyst is warned, however, that for most individual applications the configuration is a fixed parameter and even where he can justify added peripherals they might not be available until after the planned implementation of the application.

14.2.2 *Hardware*

There are a number of ways in which the storage device adopted may influence the organisation structure of a file. To take the most obvious case: if the only storage device available is a magnetic tape unit, then a sequential structure will be implied. The impact of hardware on file organisation is greater than this, however; it is, for example, vital that the interaction of the physical and logical file organisation be considered in the light of the hardware available. It is highly desirable for the overflow records in a randomly organised file to be located on the same cylinder of a disk pack as the main body of the file, to prevent read/write arm movements when the overflow area is accessed. Whether this interaction is the concern of the analyst or of a programmer will vary between installations, but in any case the analyst should be aware of the factors involved.

The concept of the cylinder on a magnetic disk unit is also important in the physical organisation of a file. By storing a file on the same track of each surface of the disk the amount of arm movement (and hence time-wasting mechanical movement) is greatly reduced. (Fig. 14.13.)

STRUCTURE	ADVANTAGES	DISADVANTAGES
Sequential	● Binary search possible on direct access storage devices. ● Fast access per relationship (i.e. if retrieval is required in same sequence). ● "Natural" method of organisation, i.e. easiest to understand. ● Security (updating by copying means previous copy of file always available). ● Minimum storage space required (space needed for pointers etc.) ● All record types supported.	● Slow searching for a particular record (on average half of the file is read). ● Updating by copying to cater for insertions means all records in file copied even if only small percentage updated. ● Only one key per record (may lead to multiple files containing same data).
Indexed Sequential	● Less handling of inactive records. ● Reduced access times for any record required in a random manner *or* if low percentage of accesses to file during run even if input and file in same sequential order.	● Updating by overlay—no automatic security. ● Use of overflow areas necessitates periodic re-organisation runs. ● Overflow pointers/indices reduce space utilisation. ● Not all record types supported by all operating systems/languages. ● Care must be taken in placing of indices to prevent excessive seek times.
Random	● Any record retrieved by single direct access (if calculation method used). ● Individual records may be updated.	● Hardware may limit record length by physical address. ● Overflow problem ● Inefficient use of storage space. ● If dictionary method used, this will need to be searched before accessing record required.

FILE ORGANISATION AND RECORD LAYOUT

STRUCTURE	ADVANTAGES	DISADVANTAGES
List/ Partially Inverted List/ Inverted List/ Ring	● Many lists possible in one file. ● Reduces need to store same data organised in different ways. ● Saves storage space.	● Space needed for pointers. ● Searching lists is effectively same process as sequential file. ● Need to keep direct dictionary of start points (list and partially inverted list structures.)

Fig. 14.12. The advantages and disadvantages of the major file organisation structures.

14.2.3 *Programming Possibility, Ease and Efficiency*

Not all file organisation systems described in this chapter or in other literature are supported by all programming languages. Even where the file organisation technique can be programmed in a programming language in use within the installation, the technical difficulty of programming, or the inefficiency of programs to process these files, may more than offset the advantages claimed.

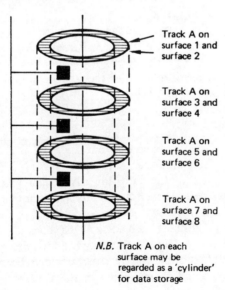

Track A on surface 1 and surface 2

Track A on surface 3 and surface 4

Track A on surface 5 and surface 6

Track A on surface 7 and surface 8

N.B. Track A on each surface may be regarded as a 'cylinder' for data storage

Fig. 14.13. A diagrammatic representation of a magnetic disk showing the cylinder concept.

If the analyst has any doubts about the technical feasibility within his installation of a file organisation system, he should discuss them with the chief programmer or other competent software advisor. It should be particularly noted that, although it may be possible to program a file organisation in a language available on the machine in use, this may not be one of the languages normally used within the installation. The analyst should only propose the use of such a structure as a last resort because of the continuous training and maintenance problems involved.

14.2.4 Operating System

This again is an area in which the advice of a software expert should be sought in any case of doubt. Suffice it to say that not all operating systems support all file organisation systems and the analyst must ensure that any organisation he proposes is feasible *within his installation.*

14.2.5 File Size

In considering the file organisation system to adopt, the analyst must take into account the estimated size of the file. Obviously the response time handicap imposed by a sequential search is much less for a small file than for a large one. Similarly for small files the space wasted by adopting a random organisation system using the calculation method is likely to prove excessive in relation to the overall file size and the benefits gained from the random organisation method.

In general it may be stated that the smaller the file the less likely it is that a complex organisation structure will be needed. Apart from this general influence of the size of a file on the organisation structure adopted, the analyst will also need to consider the inter-relationship between file size, inquiries and processing strategy.

14.2.6 Record Size

A simple record is likely to have only one key, but as a record becomes more complex it is likely that other keys will be included. These keys may not be realised at the time the file is initially designed, but may become important at a later date. If such a development is anticipated when the file is established (or if there is a reasonable likelihood that such a development will occur), it is sensible to provide a file organisation structure which allows the use of multiple keys. In this way the necessity to create files of identical data organised into different structures may be largely avoided.

Record size also should be examined in the light of the medium on which the file is to be stored. For some disk devices, fixed-length units of the file are read and to ensure the maximum efficiency of storage space this figure should be some multiple of the record length.

14.2.7 Method of Interrogation of a File

The way in which a file is interrogated will influence the organisation adopted. One is interested to know whether the enquiries arise singly or in batches, and, if in batches, whether they are in any particular order, or random. Obviously if the enquiries naturally arise in a particular order there is a *prima facie* case for organising the file sequentially in the same order. If, on the other hand, the enquiries arise in random order, but are batched for computer processing, the analyst must consider whether it would be more economic to sort the data, either off-line or on-line, before accessing the file or to use a file structure suited to the processing of random enquiries.

14.2.8 Updating Frequency and Volume

These two closely related considerations are among the most important determinants of the file organisation structure adopted. If a large volume of updating is necessary at frequent intervals, the most economic method is likely to be by making the amendments and creating a new file, leaving the old file intact. If the updating information, moreover, arises naturally in a sequential order, e.g. a payroll application, the basic data of hours worked will probably be captured and keypunched in batches which are in sequential order. Then there will be a strong case for adopting a sequential organisational structure. If, on the other hand, the data arises in random order, the analyst must consider the trade-off between the time needed to sort the data (either on-line or off-line) combined with the simplicity and economy of a sequential file structure against the time saving and added software complexity of a random file structure.

14.2.9 Response Time Requirements

As will be seen from the section on the various types of file organisation structures, one of the most important *practical* differences they offer to the analyst is the average length of time needed to retrieve an individual record from a file. This response time is in fact comprised of two or three separate elements. The first element is the time to calculate or find the key to the required record and, of course, is only applicable for some structures. The second element is the time to find the actual record required on the storage media and this may be a multi-stage process (e.g. in the indexed sequential file structure). The third element is the time needed for the physical transfer of the data from the storage media to core.

The three factors that influence the response time are, therefore: the structure of the file, the access speed of the storage device and the transfer rate of the storage device. To obtain the average response for a file time, the analyst must calculate the actual physical movement needed to bring the record to the reading mechanism of the storage device and add on the transfer time and (if appropriate) the calculation time. A timing formula will be found in the manufacturer's manual on the appropriate device.

The response time so calculated may be important in two ways. Firstly, it will affect the running time of a program operating in conventional batch mode (unless, of course, the operation is fully overlapped by other processing or peripheral operation), and, secondly, in a real-time system it will be a major determinant of the customer service provided. For example, in a real-time airline reservation system there is little, in the short run, that can be done to improve the physical operation of the storage devices supplied, as in the case of such a system these are likely to be the fastest commercially viable devices available. Furthermore, assuming reasonable efficiency of programming little can be done to improve the speed of processing the data. The only possibilities on improving the overall performance of the system are therefore in the data communications aspects (see Part III, 17) or in an alteration to the structure of the files which enables the data to be retrieved more rapidly. Thus the overall response time requirement may dictate the file structure adopted, as in the short term this is the only area in which effective alterations to the overall timing will be obtained.

14.2.10 Cost

At all stages the analyst should be aware of the cost implications of his work. In no area is this more important than that of choosing the file organisation structure, since this may well have a considerable effect on the processing time required *every time a system is run* and thus have a major effect on the running costs of the system. The set-up costs for the files and any necessary software writing to support an alternative file structure are one-time costs that must also be taken into account.

14.2.11 Space Management Considerations

Storage space is always expensive and this is particularly true when applied to data processing, and as a rule of thumb the faster the storage device the more expensive it is. It should therefore be a matter of concern to the analyst to economise on the amount of storage space used by any one file. This economy can be effected in four ways:

- Economising on the data stored in each record.
- Using data compaction techniques.
- Economising on the number of records in the file.
- Using a file structure which is economic of space required.

Economising on the data stored in each record must be considered very carefully. Obviously, where possible, repetitive data should be stored externally to the individual records in a file, but care must be taken not to exclude potentially valuable information from a file.

Data compaction techniques offer a safe and usually easy way of reducing the volume of data that has to be stored. The techniques available may be classified as: editing, the use of variable-length records, packing and encoding.

Editing is the process of removing from a record information which is used as an aid in comprehension by humans. Thus reference numbers frequently include oblique strokes and hyphens to facilitate reading, but these are quite unnecessary for machine processing and their omission can provide worthwhile space savings. Such editing marks can, of course, be automatically regenerated when output for human use is produced.

The second technique of data compaction is the use of variable-length records. A fixed-length record will almost always be the length of the longest record and in a file of records containing variable amounts of information the use of fixed-length records will mean that all records are the same length as required by the longest record. This will clearly mean an uneconomic use of the storage space. A description of the types of variable-length records will be found on page 116.

Packing is the facility which exists on many computers to increase effectively storage by compressing the form in which the data is stored. This facility varies between machines and the analyst is advised to consult the appropriate manufacturers' manuals.

Encoding techniques vary from simple abbreviations to the use of elaborate conversion techniques based on the frequency of occurrence of particular values within the data. Particularly valuable is the use of single bits as 'flags' to denote the presence (or absence) of a particular characteristic, e.g. in a personnel file the ability to speak a foreign language might be marked by the use of a single bit. Another useful technique is to use bit patterns for coding and, since many machines use 8 bits to represent 1 character, this enables 2^8 binary patterns or codes to be stored in the storage space normally used for one character.

With all these data compaction techniques, there is the possibility of increased difficulty in the programming. Some operating systems and programming languages, for example, do not readily support the use of variable-length records on all types of storage device. In case of any doubts, therefore, the analyst should discuss the matter carefully with the chief programmer or other competent person before committing himself to any of the techniques mentioned above.

Economy in the number of records held on the live file can be obtained by frequent *purging* of the master file to delete records no longer active, e.g. deleting employees who have left the organisation from a personnel file. This involves a simple trade-off between the savings obtained, i.e.

reduction in storage space and hence quicker retrieval of data, against the processing time needed for purging. It is probable in any case that it will be found necessary to retain 'dead' records on some files for year-end processing.

Some file structures require the use of additional information with each record, e.g. the list structure (page 103). Others tend to leave areas of the storage media unused, e.g. random files using a calculation system to provide the key (page 102). Before adopting such a file structure the analyst must consider whether the advantages outweigh the relatively ineffective (and, therefore, costly) use of storage space.

It is also necessary to consider that when using variable-length records and updating takes place *in situ* the updated record may well exceed the length of the 'old' record. In these circumstances care must be taken to avoid overwriting the next record and also to ensure that the updated record is stored in full. This may well require the provision of free space throughout the file or an 'overflow' area. Where overflow areas are used they should be situated to minimise the amount of extra time needed for the retrieval of records from that area, e.g. on disk additional arm movement should be avoided.

For any file structure that gives rise to overflow records or unallocated space within the file, regular space compaction runs will be necessary. This overhead must be taken into account in assessing the viability of a file organisation structure.

14.2.12 *File Security Considerations*

When designing a system, the analyst must pay particular regard to the security of that system in operation to ensure that the data it provides is always available as required and always as accurate as possible (see Part III, 15). In no case is the need for security greater than in safeguarding the existence of files and the accuracy of the data they contain.

Some physical media readily provide security, e.g. magnetic tapes are normally updated by copying (the Grandfather, Father, Son principle illustrated in Fig. 14.14) and thus a security copy of the file is generated automatically during normal processing.

For most direct access devices, however, updating is by overwriting and, therefore, it will be necessary to devise some other method of ensuring the security of the file.

The technique adopted for this purpose is that of 'dumping' the files at regular intervals. Using this technique a copy of the file is made (usually on magnetic tape) and this copy is held as security for the live file in case the latter is damaged or the data on it corrupted in any way. If such damage or corruption occurs, the dumped version of the file is used to recreate the live file by repeating all processing that has taken place since

FILE ORGANISATION AND RECORD LAYOUT   115

the copy was made. The added processing time needed for these security measures may, in some cases, offset the saving in time achieved by the use of direct access devices and may thus enable the analyst to reconsider the use of sequential files stored on magnetic tape.

	Tape A	Tape B	Tape C
1st run	New master	Security	Old master
2nd run	Old master	New master	Security
3rd run	Security	Old master	New master
4th run	New master	Security	Old master

CYCLE REPEATS

Fig. 14.14. The grandfather, father, son principle of file security.

14.3 File Organisation Terminology

In a discipline bedevilled by the use and misuse of jargon, file organisation perhaps suffers more than any other facet of the subject from the looseness with which the terminology is used. The basic terms and the concepts they represent are, however, simple.

Data structures may be likened to an inverted tree, the lowest level of this structure represents the individual data elements with which a specific value may be attached. The highest level of this structure represents the *file* of data and a number of files may be associated to form a *data base*. Fig. 14.15 illustrates the basic concepts of this structure.

14.3.1 *Data Element*

A data element is the lowest level logical unit in any data structure. It is the only level of the data structure with which a specific value may be associated. Examples of data elements might be: payroll number, part number, department number or surname. Data element is analogous with the term 'field' but is to be preferred to the latter term because it does not imply association with any physical media. The terms 'item' or 'field' are sometimes used to have an analogous meaning.

14.3.2 *Group*

The term 'group' may be used to denote two or more data elements which are logically related but do not of themselves form a complete unit of information. Examples might be: the data elements surname and Christian names together form the group 'name'; the data elements number, street, city and state together form the group 'address'.

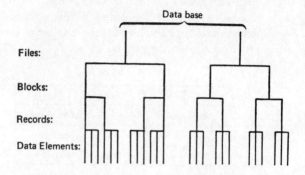

Fig. 14.15. Basic data structure terms.

14.3.3 *Record*

The term 'record' is used to denote a collection of data elements (which may or may not be associated into groups) which are all related to a common identifier. Used in this sense the definition refers to the *logical record*. A logical record may be composed of a fixed and constant number of characters in which case it is called a fixed-length record or the length of each record may vary. In this latter case it is called a variable-length record. Variable-length records may be of four basic types:

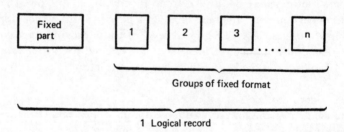

Fig. 14.16. Variable length records—type 1.

Type 1   In this case each record is composed of a fixed part and a variable number of additional data elements or groups each of which has a fixed length and format. (Fig. 14.16.)

An example of this structure might be a customer file in which the constant part (i.e. name, address, etc.) is followed by a variable number of transaction groups each of which contains in fixed format, date, description and quantity.

Type 2   Each record consists of variable numbers of fixed format data elements which may or may not already be associated into groups (Fig. 14.17).

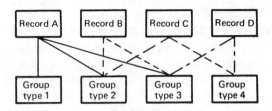

Fig. 14.17. Variable length records—type 2.

Records of this type might be found on a product file where all products are built from a range of sub-assemblies or units, some or all of which go into any product.

Type 3 Each record consists of constant data elements or groups but the length of each element or group is variable. A common example would be where the group 'name' is of variable length to allow for variations in the number of characters in people's names without a gross waste of storage space. A similar technique is often adopted with addresses. An example is shown in Fig. 14.18. The two records shown contain 23 and 47 characters respectively yet are comprised of identical data elements.

Type 4 In this structure the format of an individual record is fixed but different types of records are associated into files (q.v.). An example might be where a customer file has different record formats for cash customers, credit customers, government institutions, etc. The applications programs will not necessarily know the record type when an individual record is retrieved.

Any or all of these four categories may be permutated to form a particular structure.

INITIALS	SURNAME	NO.	STREET	TOWN	CODE	DATA ELEMENT
A	BEE	1	The Green	Acton	NW3	VALUE 1
1	3	1	9	5	4	CHARACTER COUNT 1
T R R A	SIKORSKI	1276	Wainfelin Avenue	Pontypool	NP46AQ	VALUE 2
4	8	4	16	9	6	CHARACTER COUNT 2

Fig. 14.18. Variable length records—type 3.

14.3.4 Blocks

Physically, records may be retained on the storage medium individually or in *blocks* of two or more records. (Fig. 14.19.)

The advantages of blocking records are that storage space is reduced as fewer inter-record gaps (which are required by the hardware of all magnetic storage devices) are needed and that the speed of retrieval is enhanced (less time is spent on locating records). The disadvantages are that records are always retrieved in blocks which will be unnecessary for random processing, the input area reserved in core must be increased to cater for the larger size of the unit of data physical retrieved and possibly the increased difficulty of programming.

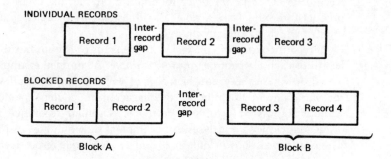

Fig. 14.19. Blocking of records.

14.3.5 Files

Related records (which may or may not be physically combined into blocks) are organised into files (sometimes called data sets). The methods in which this organisation may be effected are dealt with on page 99ff.

14.3.6 Data Base

A collection of files on a common theme, e.g. payroll personnel and labour costing files designed to avoid the duplication of data elements, is often called a data base. This term is sometimes, however, reserved for the aggregate of files to meet total corporate information requirements if designed to avoid duplication. In this latter context the term common data base or data bank is preferred.

14.3.7 Other Terms:

- Key. A key is that part (or those parts) of a record which are used for identification purposes. A key may be part of the data or a separately allocated reference number. Keys are not necessarily unique although, of course, it must be possible to identify individual

FILE ORGANISATION AND RECORD LAYOUT   119

records. With certain types of file organisation, there may be multiple keys in a record.
- Storage Devices. These are the peripheral units used for the storage of data, e.g. tape drives, disk drives, data cells, etc.
- Storage Media. This is the material on which the data is physically recorded, e.g. magnetic tape.
- Sequential Access Media. Those media which can only locate the next or previous record in sequence, e.g. magnetic tape, paper tape.
- Direct Access Media.† Those media in which a hardware address is used to position the read/write mechanism directly to the physical position where the data is to be written, updated or from which it is to be retrieved.
- Logical Organisation. The logical organisation structure of a file is the way that the file *appears to be organised to the user*. Thus a file will appear to be stored in the sequence (or in the series of sequences) specified by the relationships stored in that file.
- Physical Organisation. The physical organisation of a file is the way it is actually stored on the storage device. In some cases, e.g. a sequential file stored on a magnetic tape, this will be identical to the logical organisation. With most structures, however, the logical and physical organisations will not be the same.

† N.B. Although commonly referred to as direct access devices the method of access for disks, drums and similar devices is in fact followed only to the track. The search for an individual record or block on a track depends on the rotation of the disk or drum past the read/write head and is, therefore, sequential.

15
DATA SECURITY

As the systems analyst has the design responsibility for those systems which he proposes, he also holds the responsibility for ensuring that the overall system is such that it will safeguard the organisation's interest by preventing the loss of operational capability due to internal system faults. Furthermore, he should also ensure that the overall system is protected against the effects of hardware failure and deliberate or accidental destruction or corruption of data.

The measures taken to guard against these failures may be classified under three main headings:

Physical Security (Generic Controls).
Systems Controls (Specific Controls).
Audit Controls (Imposed Controls).

15.1 Physical Security

The physical security of data should be the subject of precise standards. There are, however, certain basic rules which should be observed by the analyst and built into every system. Some of the key points are as follows.
15.1.1 *Library Control.* This centralised control should extend to cover all program files, operating systems and utility software, as well as total systems documentation. Only in this way can it be safely assumed that in a given run the correct version of the program was used on the right data and with the correct files. A specific discussion on methods of controlling the use of files is to be found in Part III, 14. It is also, of course, imperative that the computer-generated output is given the correct physical identification, e.g. tape reel labels to ensure that correct information is used in subsequent runs. This latter point is properly part of a second major area of physical security, namely:
15.1.2 *Computer Room Procedures.* The operating standards referred to in Part III, 16, which include rules governing the labelling of output, are properly part of the complete scheme of data security as are the 'housekeeping' checks such as the routine testing of data preparation equipment and preventative maintenance of hardware.
15.1.3 *Limitation of Personnel.* It is generally sound practice to limit

access to the computer facility to the minimum number of people consistent with its operational efficiency. This policy will, in any case, reduce the likelihood of data being subject to deliberate or accidental loss or corruption. The analyst must, however, see that additional limitations are imposed when the sensitivity of the information being processed demands it. The most obvious example is the processing of company payroll information, at which time many companies have a policy that only one senior operator or perhaps wages clerk will attend the printer when salaries are printed. From the point of view of commercial security, however, many other systems will have a much higher sensitivity and precaution should (and must) be taken to see that printouts and files are not left freely accessible.

15.1.4 *Natural Hazard Protection.* This area of security covers fire, flood, storm and riot protection and should, of course, be provided as standard for all systems. Prior to implementation of a new system, however, the analyst should check that adequate safeguards in the form of duplicate programs (including supporting software), and back-up files will be created and maintained in a location remote from the data processing centre. He should also check that whatever back-up arrangements do exist (such as using a nearby installation with a similar configuration) are capable of handling the system both technically and in terms of volume. If this back-up does not exist—or cannot be guaranteed within an acceptable period of delay—alternative arrangements must be made.

15.1.5 *Hardware Controls.* All modern hardware incorporates a certain amount of checking to provide against technical errors. The best-known example of this is a 'parity check'. This detects data which has been distorted by the loss or gain of 'bits', which sometimes happens during the frequent transfers and storage of data within the hardware system.

15.2 Systems Controls

This term is used to cover all those controls that the analyst specifies for inclusion, i.e. those controls which are *specific* to a particular system. The actual controls included will, of course, depend on the sensitivity of the system, but there are a number of underlying principles to be observed *during the design phase*—for it is important that these controls be embodied in the system, not added at a later stage to pay lip service to some management or user edict. This embodiment will obviate the danger that the controls will not be applied during normal operation. These general considerations may be identified as follows :

15.2.1 *Control should be exercised as soon as possible.* The adage that raw material should be inspected to avoid rejection of the finished goods is as true of data processing as it is for manufacturing industry. This concept is neatly summarised in the acronym GIGO—'Garbage In, Garbage Out'. The best way to ensure that a data processing system does not produce erroneous results is to provide correct input. Thus the analyst should devote considerable attention to the origin of the source data and arrange controls to isolate and correct mistakes as early as possible in the chronological sequence of processing.

15.2.2 *Controls must be defined.* It is imperative that the precise manner in which each control is carried out (whether by machine or manual intervention) is identified, agreed with all parties concerned and recorded. Furthermore, responsibility for each manually executed control must be similarly identified. This responsibility is best tied to an organisational position, e.g. chief wages clerk, rather than Mr Jones, because of the confusion that can arise when Mr Jones is promoted, is transferred or leaves the organisation. The system must also include provision for identified errors to be returned to their point of origin for correction. Only in this way will it be possible for the error rate to be improved.

15.2.3 *Adequacy of Control.* A further principle to bear in mind when establishing the controls to be included in a new system is that control should be adequate but not excessive for the purposes required. Adherence to this principle will involve an estimation of the acceptable level of errors in the output and of the costs and efficiency of each method of control.

The immediate reaction to the question 'What is the acceptable level of error?' is almost invariably—'NIL', but a little reflection will soon show that this is very rarely true. Clerical systems are always subject to error because of the human factor, and this will continue to be true for a computer system as well. Moreover, there is a finite (though very small) chance of machine malfunctioning causing errors. The theory of 'zero defects' is, therefore, conceptually unsound. In any case within a normal business environment the cost of ensuring virtual freedom from error is greater than the possible benefits resulting from it. Furthermore, the concept of degree of error is vital. An error valued at £1,000 is obviously of greater importance than one of 5p although both may represent a single 'mistake'.

The aim of the analyst must always be to reduce the level and degree of errors to or slightly below the tolerable level without incurring unreasonable costs.

The other part of this evaluation, i.e. the estimation of the cost of the control, is not always easy to carry out because of the multiplicity of cost elements involved. Some estimation is, however, necessary if the cost-effectiveness of additional controls is to be established. The major cost elements will be:

- The time to establish and maintain the control.
- The time to administer the control.

- Additional computer time.
- The cost of the additional delay imposed on processing.
- Special hardware (or increased hardware).
- Material costs (stationery, etc.).

The ultimate decision on the acceptable level of control (or of error) should not be made by the analyst. Indeed, the responsibility for this decision is not within data processing at all—it is a business risk decision which must be made by line management of the user area(s) involved. The analyst does have a particular responsibility, of course, to ensure that the risks and the cost of their elimination are clearly presented to management and to use all the techniques available to minimise both risks and costs.

15.2.4 *Error Statistics.* Any major system should include provision for statistics of errors to be maintained. This will enable the effectiveness of each control to be ascertained and allow analysis to show which errors could have been identified and corrected earlier in the system. Not only will this enable subsequent improvements to be made to the system but will also enable more effective controls to be incorporated in other systems.

Having identified the general considerations applicable to systems controls, the question arises what specific controls may be incorporated in a particular system? The following list identifies the major systems controls and identifies the parts of the total system at which they may be applied.

(1) *Control totals*

The principle of control totals is that the value of a particular data element is summed as a separate clerical process and when the input documents containing the same data elements are processed by computer a machine-generated total is created and reconciled to the control total. This system is mainly used to control input for batch process systems.

Two closely similar controls are also in common use. These are:

- Hash totals. Hash totals are basically the same as control totals except that the total itself has no intrinsic value. Thus a total of the data element 'gross pay' on a batch of input documents for a wages application would be a control total; the total of payroll numbers would be a hash total.
- Document counts. With this form of control only the number of documents is summed, not any particular value contained on those documents.

(2) *Supervisory checks*

Checks of this nature are, of course, implied, if not stated, in all systems specifications. It is a basic responsibility of any level of management to monitor the work of subordinates. The analyst is, however, ill advised to rely on this procedure to provide any system control, since it is subject to all the vagaries of human nature, not to mention absence or lack of attention due to pressure of other work. Checks of this nature should be seen as supplementary or as a means of enforcing other controls—not as controls in their own right.

(3) *Reasonableness checks*

This represents a wide range of control techniques of both manual and machine types. The basic principle is that the data is examined to see if the value is a *reasonable* value for that data item. Thus, under normal circumstances, a net pay figure of £17 per week would in most cases be reasonable but one of £900 per week would not. When used as a manual control, this is only a formalisation of the 'Does it look right?' attitude subconsciously used by all good clerical staff. Such checks can, however, be written into computer programs both to validate input (as in data validation runs) and also to check intermediate and final results.

When checks of this type are used, care must be taken as to what action is to be taken when the check reveals an 'unreasonable' value. If this results in the rejection of the result and subsequent examination reveals that *in this instance* this 'unreasonable' value is valid, a method must exist to ensure the production of the final output. If, for example, a reasonableness check on a payroll application is set at £99 per week and due to some abnormal circumstances, e.g. the payment of a back-dated rise, results in one or more persons legitimately exceeding this figure, provision must exist for producing payslips in these cases.

(4) *Check digits*

These are also called terminal check digits and modulus n checks. This is again a technique which may be employed at two or more stages in a total system. The basic principle is that a reference number, e.g. payroll or customer number, has a supplementary number added to it. The supplementary number which is usually suffixed is calculated to bear a definite arithmetic relationship to the rest of the number. The complete number can then be checked by performing the same calculation on the root number and comparing this to the previously computed suffix. If there is agreement, the number is validated; if not, the number is rejected. Some types of data preparation equipment are equipped to perform this calculation and it can also be used in computer programs to validate input, such as part numbers. One means of calculating a check digit for modulus 11 (probably the most popular in general use) is shown in Fig. 15.1.

(5) *Verification*

It is important to realise that the decision to verify (or not to verify) data input is part of a study of a total series of system controls. Statistics are available to identify the residual error after verification under normal circumstances. Normal rates are in the range .02–.05%. These figures are, however, concerned only with the accuracy of the punching (taping or disking) from the source document—not the accuracy of the information on the source document.

Calculation of modulus 11 check digit

Basic part number =	1 9 8 7
Each digit is multiplied by a digit corresponding to its scalar position	×5 ×4 ×3 ×2
Add products	5 + 36 + 24 + 14 = 79
Divide products total by modulus	$\frac{79}{11}$ = 7, remainder 2
Remainder is subtracted from modulus to give check digit	11 − 2 = 9
Check digit =	9
Therefore full part number = (with check digit)	19879

N.B. Where check digit = 10 (highest value possible using modulus 11), the letter 'A' is normally used as the check digit

Fig. 15.1. Calculation of modulus 11 check digit.

(6) *Absolute checks*
The absolute checks are a comparison between the actual value of a data element and the valid values which can be attributed to that field. Thus it is only really applicable to reference information, e.g. part numbers, etc. Although conceptually such checks can be applied, manually their application is limited in most cases by the volume involved with computer use. Their limitation is the practical difficulty in ensuring that the master file, against which the data is checked, is updated to cater for all valid input values.

(7) *Data sequence checks*
Certain data elements must occur in specific sequences and these sequences may be checked by program. Thus, in a simple example, transaction data in a card-oriented application are only valid if preceded by a customer data card with the same reference. Where such sequences naturally exist in data, they should be checked by program.

(8) *Identify retention and validity dates for files*
The data included in each file must be clearly identified both physically— to ensure correct manual handling—and also internally. The internal identification will vary according to the media used, being unitary for files composed of discrete units (e.g. punched cards)—see also page 85—and

taking the form of header and trailer labels for files on continuous media, e.g. magnetic tape. Program specifications must include routines for checking the labels, thus ensuring the processing of the correct data. The contents of a typical header label would include:

- File name.
- Reel or disk number.
- File generation number.
- Retention period.
- Date file was written.

Header labels should also be checked when writing files to ensure that existing data on that particular tape or disk has exceeded its retention period and may safely be overwritten.

(9) Character or picture checks

Checks of this nature are a useful means of program checking of input data without recourse to an absolute check. The principle is that many individual data elements have a fixed format and this can be compared with a stored 'picture' of that format. For example, part numbers may be allocated from the range AA0001A–ZZ9999Z; thus any value of data element having the format 2 alphabetic characters, 4 numeric characters, 1 alphabetic character may be valid. The adoption of this type of check is not limited to those circumstances where the format is completely fixed. If, in the above example, the second character could be either numeric or alphabetic, a check could still be performed to eliminate blanks or special characters.

(10) Sampling checks

It is often useful, especially in the manual areas of a total system, to institute a check on only a proportion of the total data volume. The type of check may in fact be any of those outlined above—the important point being that, when sample checks are used, only a fraction of the total data is checked. Thus, for example, 10% of the input documents to an inventory system may be subjected to an absolute check on the data element 'part number'. If the percentage of errors is higher than an agreed tolerance level, checks on a higher proportion of the volume—perhaps even 100%—may be instituted until the cause of the error is identified and corrected.

These are the major principles and controls which may be incorporated into a system. There are, however, some additional points that are worth remembering during the design and implementation phases which can significantly reduce the number of errors in a system. These are:

- Good source documents will reduce data preparation errors—see Part III, 14.

- Good staff training and user-oriented documentation will minimise errors caused by incorrect completion of source documents or misunderstanding of output.
- Standards will minimise errors through misunderstandings, e.g. does O mean zero or the letter O?
- The smaller the volume of input into a system the fewer the number of errors—it follows that the analyst should use every possible means to reduce the input volume, e.g. use data already in machine-readable form whenever possible, derive data by machine rather than manually, etc.
- Keep input batch sizes for data preparation as large as possible. Tests show that in general larger batch sizes result in a smaller percentage of errors because the operators develop a rhythm peculiar to the work.
- Promote awareness of other people's problems. If staff are acquainted with the work (and problems) of other staff who occupy a position later in the chronological chain of processing, they will in general be more careful in their work.

15.3 Audit Controls

In addition to the controls which are developed for system integrity in response to the needs of the user and the judgement of the analyst, there is another category of imposed controls which are designed to meet legal and audit requirements. It is important to realise that these requirements do not only apply to those systems which involve money. Any system which involves stocks or raw materials (and this will include production recording) will also be of interest to auditors. The basic requirement in this connection is the provision of a trail—'The audit trail'—which enables the calculations performed on a certain item of data to be traced back through each processing step to source.

Legal requirements which are also in the same category of imposed requirements will vary not only with location but also with time. The analyst is reminded that legal requirements may exist for such apparently innocuous systems as safety (or accident) records or transport vehicle utilisation records. Whenever the analyst is designing a system which may have imposed controls for audit or legal requirements, it is necessary to consult the company auditors and/or lawyers at the earliest possible opportunity. They *should be involved in the design process* or at the very least should carefully vet the proposed design *before implementation*.

16
STANDARDS AND DOCUMENTATION

Data processing today represents a major investment for most companies. This investment is measured not only by the capital involved but also in terms of the time and effort devoted to it. As in any other area of the company, business control must be exercised over this investment and standards provide the yardstick which can be used for this purpose. By 'standards' we mean *the use of uniform practices and common techniques, which provide a basis for assessing the performance in terms of both quality and quantity.*

Standards also have a major role to play in the training of staff and the desperate shortage of data processing staff can be at least partially alleviated by the use of standards. Training of new recruits to data processing is eased if there are comprehensive standards in use, as the trainee can become productive in a shorter time if he is following established procedures than if he has to develop his own. Retraining experienced staff recruited from other companies is also aided by the existence of standards —especially if they are common to both his 'old' and 'new' employers. Furthermore, the use of good standards can in many cases mean that a task can be carried out by a less experienced person than would otherwise be necessary, e.g. if good standard documentation is available, program maintenance can be undertaken by less experienced programmers than when it is necessary to decipher the logic from a program listing.

Standards also alleviate communications problems and reduce the dependence on individuals. This means that less disturbance is caused by holiday and sickness. Moreover, the company cannot be held to ransom by an individual who is the only person with a detailed knowledge of a particular system. In alleviating communications problems, standards are instrumental in raising the overall level of efficiency within the data processing field and increase the level of professionalism among the staff. In the field of systems analysis standards enable full use to be made of previous work in the same area and ensure that vital points are not missed. In programming the use of standards means that *one of the better techniques* is used in the solution of a problem. That an individual programmer *might* come up with a better solution in a particular set of circumstances is not disputed, but any saving achieved over the adoption of a 'standard' routine must be balanced against the time (and therefore cost) of developing the alternative and thereafter maintaining it.

In the operating area the case for standards is perhaps the most obvious of all. A computer operations team can be likened to any other production facility and clear, concise instructions given for day-to-day operation. To reduce the possibility of misunderstandings and omissions, these instructions are best presented in a standard manner. The use of performance standards in this area, e.g. utilisation of the central processing unit, will also be appreciated.

With the increasing sophistication of data processing systems, the days when one analyst/programmer could perform all the steps necessary to computerise a system are well-nigh over. This and the increasing integration of systems means that many people will be working on different areas of the same problem. This presents a particular kind of communications problem if time-consuming repetition of work is to be avoided and at the same time all the work is to be capable of being formed into a coherent whole.

Standards are particularly valuable when conversion to a new computer takes place (see also Part IV, 20 and 21). Firstly, the existence of standards enables a more realistic appraisal to be made of the work entailed than would otherwise be the case. Secondly, the use of a standard programming language, e.g. ANSI COBOL, will reduce the amount of re-programming needed. Thirdly, where re-programming is necessary the existence of standard documentation and the use of standard routines will significantly reduce the amount of work needed.

A number of reasons against the introduction of standards are sometimes quoted and, in spite of the strength of the case for standards, it is therefore necessary to examine these before concluding that standards are a definite benefit to data processing.

Some people claim that standards restrict creativity. There is a limited amount of truth in this argument. It is only valid if the exercise of creativity would produce a better solution to the problem. It follows that, provided the standards are created by the most creative people and that they are revised as better techniques are developed, the benefits of creativity are in general realised whilst the benefits of standards are retained.

A second argument against standards is that they increase the workload. This is probably true in the short term but it is certainly false in the long term. A certain amount of time (around 5%) is needed, for example, to produce standard documentation, but there is a longer-term saving far in excess of 5% because maintenance and conversions are both greatly eased by the use of standard methods and the existence of standard documentation.

It is also argued that standards can slow the development of a system because the existence of a formal set of procedures and documentation does restrict the pace at which these procedures may be altered. This is in reality the case of anarchy against order and the cause of the installation is almost certainly best served by bringing the weakest systems and personnel up to the agreed minimum level by the imposition of standards even if this does mean some restriction on the most able or advanced.

One problem which may arise if standards are wrongly applied is that they encourage people to work down to the standards rather than raise their level of performance to the highest possible level. There are two answers to this problem. Firstly, the standards must be set at a level to encourage people to work up to (particularly important for performance standards) and secondly that this problem is caused, not by the existence of the standard, but by the failure of the management involved to convince the staff of the need for standards. This topic is discussed further in the section on the introduction of standards.

It may be concluded, therefore, that on balance there is far more to be said for adopting a full set of standards than can be stated against their adoption. Perhaps, however, the biggest single reason for adopting standards is that they demonstrably improve the professionalism of data processing. Gone are the days when the computer people could afford to be regarded as mild eccentrics on the fringe of a company's activities. The computer and the people who work with it are now an integral part of most medium and large companies and increasingly playing a part in small companies, too. They must therefore be, and be seen to be, subject to at least the same degree of departmental self-discipline as any other professional group in the company.

Even when the need for standards is recognised, however, there is a tendency for a few rules to be drawn up for program documentation and to call these a set of standards. Although documentation is indeed an important part of a total set of standards, it is by no means the complete story. To secure the maximum advantage, they should cover all areas of activity within the ADP function, i.e. analysis, programming and operations.

Systems analysis is usually the area within the ADP activity in which least attempt has been made to introduce standards and yet, because of its position as the source of work for programming and ultimately operations, the case for standards is probably even greater than in other areas. Standards can be applied to systems analysis in a number of ways, the most obvious being the use of standard documentation, particularly for the production of systems and program specifications.

A standards approach to systems analysis can, however, extend further than this and provide a framework for the performance of the task. Part I of this book describes a standardised methodology for systems analysis, which would also assist in training new staff and improving efficiency of existing staff.

Within an overall standardised methodology of systems analysis, a standardised approach can also be adopted for fact finding through the use of procedure description sheets and report form documents (see page 139ff). The use of such aids has three major benefits. Firstly, they ensure that important facts are not omitted or forgotten; secondly, they provide a means of communications between persons working on the same job; and thirdly, they facilitate the subsequent maintenance of the system as either the original or other analyst can readily ascertain the relevant information which leads to the introduction of the system.

16.1 Performance Standards

The definition of standards quoted at the beginning of this chapter includes reference to the fact that standards provide yardsticks against which performance can be measured for quality and quantity. From the areas mentioned above the use of standards as a yardstick for quality can easily be appreciated, e.g. does this program specification meet the standards laid down for documentation? It is perhaps less easy to see how standards can be used to ensure the quantity of performance except in those cases where performance is easily directly related to some definite machine function, e.g. number of keystrokes performed by a keypunch operator.

Moreover, the knowledge that standards are being used as a quantitative measure of performance may lead to some problems with the staff involved unless they clearly recognise the need for some such measure. The major reason for the introduction of performance standards is, however, not the direct measurement of staff work but to provide for management planning purposes a reliable means of estimating the time needed to carry out a particular task or series of tasks. As such, the introduction of performance standards is a highly desirable aid to planning (see Part IV, 21).

Reliable performance standards for some aspects of data processing work are fairly easily derived using conventional work study techniques, and these can be perfectly satisfactory for data preparation and similar activities.

Performance standards for these aspects of data processing work which have a significant creative content, e.g. programming and systems analysis, are less easily obtained, but a number of techniques are available at least for program writing which enable consistently reliable estimates of workload to be developed on an objective basis. These techniques are generally based on a weighting system for all the relevant factors, programmer experience, test opportunities, etc., applied to a fairly rough assessment of the size of the program.

Within the field of system analysis the subject of performance standards is at a less satisfactory stage of development. Some techniques have been tried by allowing a percentage of the estimated programming time for the system as the time necessary, but in the main these suffer from the defect that they rely on some parameter, the value of which is unknown at the time that the estimation of the systems analysis effort would be of any use.

The present state of the art is that the subjective estimate of an experienced person is likely to be the most reliable estimate.

16.2 Sources of Standards

Once it has been agreed that a set of standards should be introduced and that they should cover agreed areas of activity, the next step towards their introduction should be to analyse the sources from which they may be obtained and to build up from those sources the standards which are most suitable for the installation. The main sources of standards are :

- Standards organisations and national bodies.
- User groups and other users.
- Manufacturers.
- Published work.
- Informal standards already in use.
- Departmental staff.

Standards organisations and national bodies include such organisations as the International Standards Organisation (ISO), The British Standards Institution (BSI), The United States of America National Standards Institute (ANSI) and The National Computing Centre (NCC). Each of these bodies (and their counterparts in many parts of the world) has been active in establishing standards. In most cases these are complementary, e.g. standards for COBOL were developed by the ANSI and supported worldwide, whereas similar work on the ALGOL language was carried out in Europe by the European Computer Manufacturers' Association (ECMA) and later by the ISO.

It is to be regretted, however, that in some instances these bodies have recommended different standards. Thus the ISO recommendation on flowchart symbols published in March 1969 was followed by the NCC systems documentation manual which recommends different standards.

User groups, which are often established with the active support of the manufacturer concerned, are set up as a forum at which users of similar equipment or machines from the same range can compare experiences and discuss common problems. As such, they provide a very profitable source of material for inclusion in a comprehensive set of standards. Material from this source may often be supplemented by informal discussions with other users, especially since in many areas a degree of informal 'freemasonry' exists between the staffs of different data processing installations. *Computer manufacturers* are another source of material for standards. In addition to their activities in promoting user groups, they are also active in the formal sphere (recommendations contained in or implied by the use of manuals and other documentation supplied to the user and also in the material presented on training courses) and in a less formal manner by the actions and recommendations made by systems engineers and other personnel working in close contact with the user.

Published work. Much good material for incorporation into a total set of standards may also be found in published work (both magazines and books). Standards drawn from such sources tend to be of a more general nature than those derived from sources more directly influenced by the manufacturers. A selection of works giving useful material in this category will be found in the bibliography.

Informal standards. In any data processing organisation it will be found that certain 'rules' are obeyed even if there is no formal set of standards in existence. These 'rules' are often established by consent of at least a majority of the staff and being demonstrably beneficial are quickly assimilated into the routine of the department. Standards which have developed in this way are worthwhile formalising into the manual of standards because they have naturally arisen from the requirements of the work and, moreover, they have the support of the staff without the need for the usefulness of these standards to be proved.

Departmental staff. The ideas of staff are another fruitful source of standards. It will be found that many ideas on what needs to be included in a set of standards and indeed in many cases recommendations for specific standards are already formulated in the minds of staff before any attempt to introduce formal standards. As in all areas of management, the views of the staff actually performing the work should be carefully considered (if not always slavishly adopted). Adopting standards from this source does, of course, reduce the number of problems that may be encountered when they are introduced operationally, since they already enjoy a certain measure of staff support.

16.3 Introduction and Enforcement of Standards

To ensure that the benefits claimed for standards are in fact realised, care must be taken over the way in which they are introduced within the concern. It is not enough to create a mass of standards and then issue them with the instruction that the standards will be applied from a specified date.

As mentioned above, the staff of a data processing department is a fruitful source of ideas for standards, and this in itself is a major reason why all staff should be involved to at least some extent in the creation of a standards manual. Furthermore, the problems associated with the introduction of standards will be largely avoided if the staff are involved in the project from the outset.

The most satisfactory way in which a set of standards can be introduced is by the issue of a *manual of standards*. This manual should cover all areas of data processing activity, including systems analysis, and must be available to every member of the staff. For ease of updating and addition, it should obey the rules of good documentation given later in this section. The manual should clearly indicate which standards are *mandatory* and which are *advisory*.

After the necessary preliminary action has been taken to convince all the members of the staff of the need for standards, they should be invited and encouraged to submit suggestions both as to the areas which the standards should cover and also the actual standards that should be incorporated. This phase should be, as far as possible, the subject of open debate within the department and ample opportunity should be given to *all staff* to express their views both on the standards applicable to their own section and also to other sections of the department. When some general consensus of opinion has been obtained, a draft of the standards manual should be issued and formal comments invited. It may be more convenient to issue the draft section by section rather than as a complete entity.

When the comments have been received and discussed, the manual should be formally issued to all members of the staff with full instructions as to when the standards become operative and a short covering letter restating the main advantages sought by the introduction of standards and a request for the full co-operation of the staff in their application.

The actual collection of standards and preparation of the standards manual is best entrusted to a small working party which represents all sections of the department. The persons serving on this working party should be selected on the basis of the professional respect they command from their colleagues.

The introduction of standards to a data processing department is an exercise in human relations that demands management of a high quality, if the full benefits are to be realised. The following general principles are intended to serve as a guide to the major areas to which management should devote attention before and during the implementation of a manual of standards :

All personnel must understand the reasons for standards and the methods being used to establish them. This is clearly a pre-requisite of any action which requires so much co-operation to ensure its success as does the introduction of data processing standards. Such understanding of the underlying reasons for the introduction of standards can be promoted in a number of ways amongst the most important of which is the developing of the realisation of present shortcomings. This is best done by the 'How do we overcome this problem?' type of management rather than the 'We must' type of management. The use of this approach can create the realisation at all levels of the need for standards which is the ideal atmosphere for their introduction.

The second way in which personnel may be convinced of the need for standards is to stress the increased professionalism it will bring to their work. Data processing is a relatively young discipline and is still in the early stages of developing its own professional standards of education, etc. The discipline of working to a set of standards is a very important way of raising the status of the discipline *vis-à-vis* other disciplines.

Spreading an understanding of the methods used to implement standards can probably best be achieved by a series of departmental staff meetings at which an important point is made of the value attached to suggestions made by the staff of the department. This technique can also be used to promote the case for standards.

It is important to stress that these activities to ensure support for standards should be continued throughout the introduction phase, and indeed afterwards, to ensure complete success.

16.4 Standards must enjoy management support

The concept of standards must be fully supported not only by data processing management but also by the management to which the data processing manager reports if success is to be achieved. This is, of course, a fundamental precept of management that you can only convince others of something in which you yourself believe and this is as true for the introduction of standards as for any other management decision.

It is also important for management to realise fully implications of the introduction of standards. As indicated above, there will be instances where the introduction of standards imposes an additional workload and this will, of course, be (at least in the short term) reflected in the throughput of the department, especially in terms of newly developed work. Furthermore, there may be instances where this effect is highly significant, e.g. in missing target dates. It is in this situation where management support for standards is really tested, and it is important that standards are enforced at these times even if their being ignored would mean a target date were met. Immediately 'one' exception is made the need for others will become apparent, and before long the whole concept will be undermined. This is not to say that standards must be applied slavishly without regard to the overall needs of the organisation, but rather to stress their importance and to ensure that full support is forthcoming from all levels of management.

It follows from the preceding discussion that management must be prepared to *enforce the standards set*. As indicated above, standards may be either *mandatory* or *advisory*. Clearly, if the standards are to have any credibility at all, the mandatory standards must be enforced and this implies that the necessary authority to do this is delegated to the appropriate persons. Thus a chief programmer should be given the authority to reject program specifications that do not meet the mandatory standards and similarly the operations controller must have the authority to reject any job submitted for either test or routine running if it is not accompanied by the appropriate documentation as specified in the standards manual.

The application of advisory standards requires a more sophisticated approach. These standards are established as desirable guide lines, but not as rigid limits, e.g. for purposes of back-up availability or running efficiency (especially in a multi-programming environment) it may be advisable to specify limits on the number of peripheral devices used or core occupied which are below the actual availability (even allowing for operating system requirements). These are properly advisory standards which can be ignored on occasion for overall systems efficiency. In such cases it should be stated in the standard that it need not apply if prior agreement is reached between all interested parties and a formal record of this agreement should be included in the job documentation.

Standards must be enforced

Like any other system of working, standards must be enforced by management action if they are not to fall into disrepute. The action taken will, of course, depend on the circumstances in any particular case, but it is, as indicated above, vital that the necessary authority is delegated to those persons who are in a position to ascertain the compliance or non-compliance with the standards as a routine decision. Provided that all staff are convinced of the need for standards and have adequate access to the manual in which they are contained, little problem should be encountered in their enforcement. The occasional deviation from the standards can then be dealt with by a gentle reminder. Persistent failure to obey standards is usually indicative of a failure on the part of management to convince the staff of their value and would indicate the urgent need for an educative process. If the failure to comply is by only one person or a small group of persons within the department it may be necessary in the end to take conventional disciplinary action against the offender(s), even if he or they are key members of the department, and management must be prepared to take this action if needed.

The standards adopted will, of course, be subject to amendment and addition after their introduction, and it is highly desirable that this function should be introduced on a sensible basis at the time of their introduction. The working party created to establish and implement the standards may well be in the best position to undertake this responsibility. The formal procedure adopted should provide for this working party to review periodically the external sources of standards and to indicate modifications and additions and also review suggestions arising within the department. These suggestions should be made formally (i.e. in writing) and, where they are not adopted, reasons for their rejection should be given. By adopting this procedure the standards will remain valid and therefore useful rather than falling into disrepute through obsolescence. When additions and amendments are made, a formal dated amendment to the manual must be issued to ensure the uniform introduction of the revision.

It will have been observed from the preceding material in this section that one of the cornerstones of a good system of standards is the creation of good documentation, both at the standards manual level, to ensure the publication of the standards themselves, and, secondly, at the operational level, as it is through documentation that many of the standards will be applied. Indeed, there is considerable truth in the statement :

'Standards through Documentation'.

There are a number of underlying principles common to all good documentation systems and particular care should be taken to comply with these points when introducing standards :

- All documentation must be clearly identified and dated. This identification should include a title, reference number, the name of the author, sheet number and date of preparation/issue. It is only by recording these facts on every document that complete job histories can be compiled. The issue date is an essential part of this reference because it is only by the addition of a date that all concerned know that they have the most recent edition of the document. This objective can be partially met by the use of an issue suffix to the reference number. This latter technique does not, however, enable an immediate reference to the time at which a particular document was current when historic searches are made, and in many instances, e.g. for audit requirements, this may be necessary information. To avoid the necessity of a cross reference to the calendar date, the inclusion of the calendar date is sufficient, and this method is preferred.

    To facilitate easy reference to filed documents, it is advisable for this information to be contained in one area of each document and for the same reason it is suggested that this should be in the same area of each different type of document. Engineering practice is to identify all drawings by a box in the bottom right-hand corner. For filing purposes this identification box should be on the right-hand edge of the form, and because most people refer to documents by turning over the upper edges of the form, the top right-hand corner is preferred.

- Each document should deal with only one topic, e.g. one print layout. In the short run this requirement may mean the creation of more documents than may seem necessary, but the justification will quickly be realised when amendment takes place. Furthermore, adherence to this requirement will enable specific documents (especially those related to data elements, see page 143) to be reproduced and used as elements in the documentation of systems other than that for which they were originally created. This concept of unit documents being collected to form total systems documentation is essentially similar to the concept of modular programs.

- Documentation should be a natural by-product of systems analysis, design and programming, etc. As far as possible, systems documentation should be based on useful working documents in order that the additional workload is reduced to a minimum. Furthermore, observation of this principle will ease the acceptance of standard documentation. The maxim that the best standards are those that are easier to observe than to ignore is particularly appropriate in this connection.

- All documents must be readily understandable and self-supporting. Each document must contain in addition to its identity either a complete key to the contents (i.e. all codes and conventions used) or a cross reference to the appropriate document. It is desirable to pre-print on every document a statement to the effect that, unless otherwise stated, all conventions followed are those contained in the standards operative at the date of issue. In this way an analyst searching through historic systems documentation is reminded that the convention used, e.g. in flowcharting, are not necessarily those currently in use. All documents must be prepared with a view to their use by other persons. Only by following this principle will documentation be produced which meets the requirements for training other staff and being readily assimilated by any other person who needs to become familiar with its contents, e.g. for maintenance purposes.

- The handling of documentation must be formalised. In the same way that standards are, in many cases, applied through documentation, it is necessary for documentation to be itself subject to standards. In addition to the principles of identification, etc., this will also apply to the handling of documentation. Among the points which may be stressed in this connection are : the maintenance of a full set of documentation for each system within the data processing library from which documents should ideally not be released and, if they are, only when a signed receipt is made, and, secondly, the provision of a duplicate set of documentation stored away from the main library to ensure protection against fire and other catastrophe (see also Part III, 15 on systems security).

- All documents must follow the principles of good forms design. In addition to the general points about identification made above, all documents should obey the rules of good forms design. If possible the advice of a specialist forms designer or experienced organisation and methods practitioner should be sought, but the following general points should in any case be observed :
  (1) An adequate filing margin should be included on each document.
  (2) Standard sizes of paper should be used. This not only eases the problems of handling, but also greatly facilitates reproduction.
  (3) Instructions for completion should be clear and unambiguous.
  (4) Adequate space should be available for all entries.
  (5) Each document should be designed with the method of completion in mind, e.g. horizontal spacing per character should be 1/10 or 1/12 of an inch for typewriters or approximately 1/8 of an inch for handwriting.
  (6) Each document should be suitable for reproduction, e.g. lines which are to be copied should be printed in a colour which reproduces well on all types of equipment. This requirement may not be fulfilled by the forms supplied by the manufacturers, e.g. printer layout sheets. In these cases the use of specially designed 'in house' forms which do meet these requirements is to be recommended. The documentation needed in systems analysis will require the use of a number of forms. Notes on the nature, use and completion of these forms are given below.

(a) *The Procedure Description Sheet* (Fig. 16.1) is a valuable aid in the fact-gathering stage of any systems analysis exercise. Its use is explained on page 48.

(b) *The Standard Report Form* (Fig. 16.2) is a general-purpose form which can be used for all narrative descriptions, calculations and diagrams and is so designed that it may be readily completed either by hand or by typewriter. The use of such a form (which can be conveniently issued to all staff in pad form in place of memo or other pads) will ensure that all documentation is created in a manner that conforms with the standards.

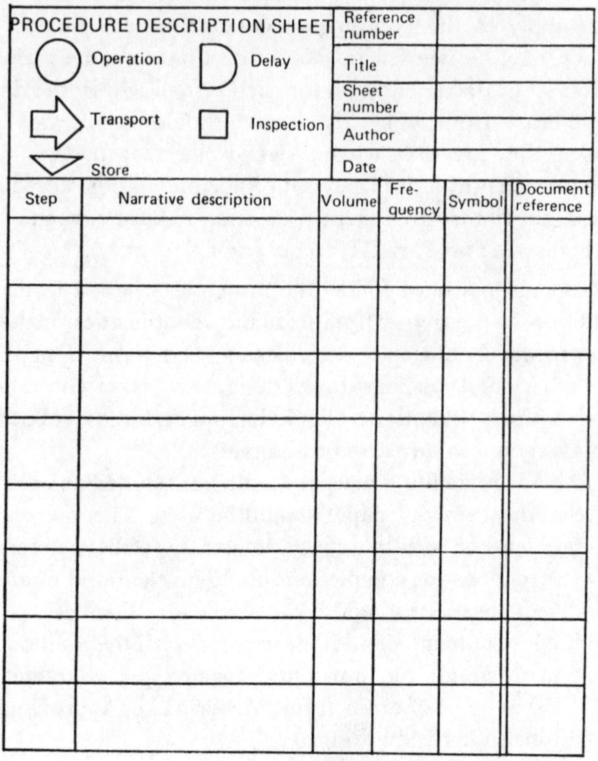

PROCEDURE DESCRIPTION SHEET		Reference number	
◯ Operation ◿ Delay		Title	
		Sheet number	
⇨ Transport ☐ Inspection		Author	
▽ Store		Date	

Step	Narrative description	Volume	Frequency	Symbol	Document reference

Fig. 16.1. Procedure description sheet.

Moreover, the use of such a form will enable many documents created as working notes, e.g. calculations of file size, to be included directly in the systems documentation without the need for a transcription process, thus promoting overall efficiency (provided always that the original document meets certain minimum legibility standards).

This form can conveniently be printed in black with a pale blue grid of lines. This will greatly facilitate completion by hand, especially when used for tables of figures, and yet will not be obtrusive and will be almost completely lost during the reproduction process by most copying techniques.

(c) *The Flowcharting Form* (Fig. 16.3) is designed to ease both the preparation of all levels of flowcharts (by the provision of an outline grid) and their reproduction (it is produced in the ISO standard A3 format). Full details of flowcharting techniques will be found in Part II, 10.

(d) *The Decision Table Form* (Fig. 16.4) provides a convenient preprinted form which may be used in the preparation of decision tables (see Part II, 11).

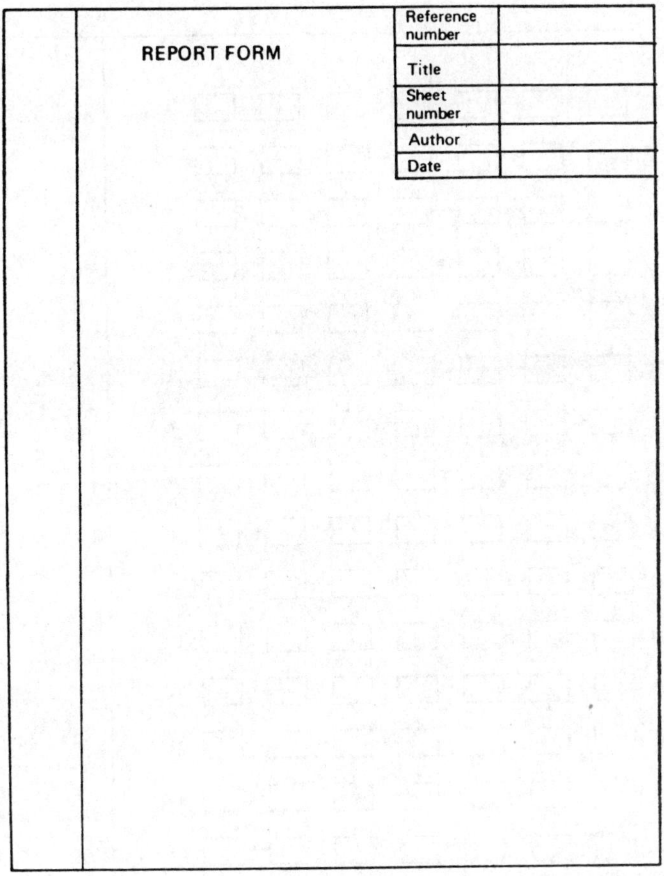

STANDARDS AND DOCUMENTATION 141

	REPORT FORM	Reference number	
		Title	
		Sheet number	
		Author	
		Date	

Fig. 16.2. Report form.

(e) *The Punched Card Layout Form* (Fig. 16.5) is basically a conventional punched card layout form modified to comply with the same standards as other documentation in this book. It will be noted that although the form is designed for completion with the longer edge horizontal the identity box remains in the normal position. This greatly facilitates reference when documents of different types are filed together.

It will also be noted that provision is made for reference to the data element description (see below) for each field of the card.

(f) *The Printer Layout Chart* (Fig. 16.6) is similar to the manufacturers' layout charts but modified to meet the same standards as all other documentation in this book. The inclusion of a facsimile of the printer control tape will assist in the specification of this item where necessary.

142 A HANDBOOK OF SYSTEMS ANALYSIS

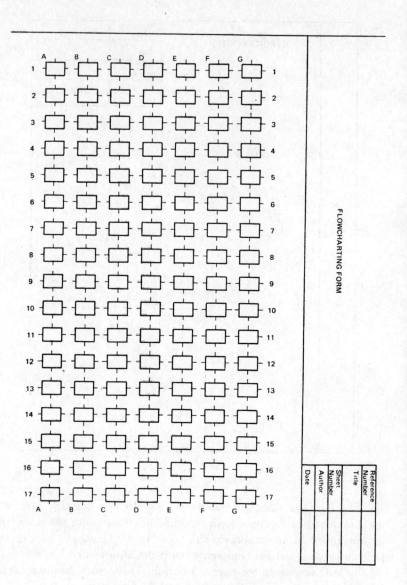

Fig. 16.3. Flowcharting form.

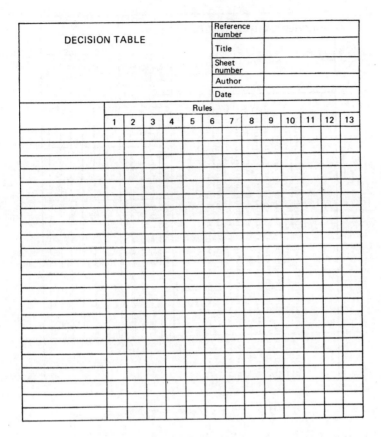

Fig. 16.4. Decision table form.

(g) *The File Specification Sheet* (Fig. 16.7). This form is suitable for designing layouts for paper and magnetic tape, data cells, magnetic disks and drums. Like the punched card layout form, reference is made to the data element description for each data element included in the record.

In using the form for any magnetic media, it must be specified whether the layout is in terms of bits or bytes, binary or decimal.

(h) *Data Element Description* (Fig. 16.8). This form is used to identify describe and record the use of each data element (q.v.). This function, already important (but largely ignored), will demand much greater attention as more and more large data banks are commissioned. It is only by maintaining close control of the data within such a system that their full potential will be realised.

The type of data element description illustrated in Fig. 16.8 will also prove a valuable aid in systems design and maintenance.

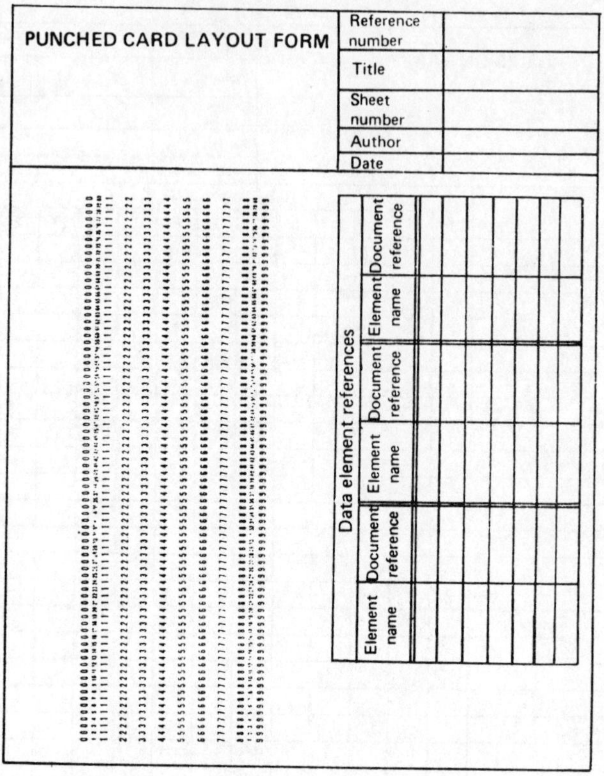

PUNCHED CARD LAYOUT FORM

Reference number	
Title	
Sheet number	
Author	
Date	

Data element references

Element name	Document reference	Element name	Document reference	Element name	Document reference

Fig. 16.5. Punched card layout form.

The use of such a form is best illustrated by an example. The completed form shown in Fig. 16.10 is for the data element 'Part Number'. Examination of the form reveals that this is a piece of basic data, i.e. an identifying code; other examples of basic data would be payroll number, employee name, customer number, etc. The other descriptions in this segment of the form are for constant, variable or resultant data. Constant data (or constants) are discount rates, pay rates, selling price or other data which (at least in the short term) have a constant value. Variable (or transaction) data are, on the other hand, data which vary with each transaction, e.g. hours worked, quantity purchased, etc. Resultant data are data which are calculated by processing, e.g. gross pay, net pay, invoice total, etc.

A unique name is allocated to each data element and it is strongly recommended that this is selected obeying the rules for labels in the programming language(s) used in the installation, so that this name can be used by programmers, thus facilitating the interpretation of program listings. This obviously necessitates the control of the allocation of such labels

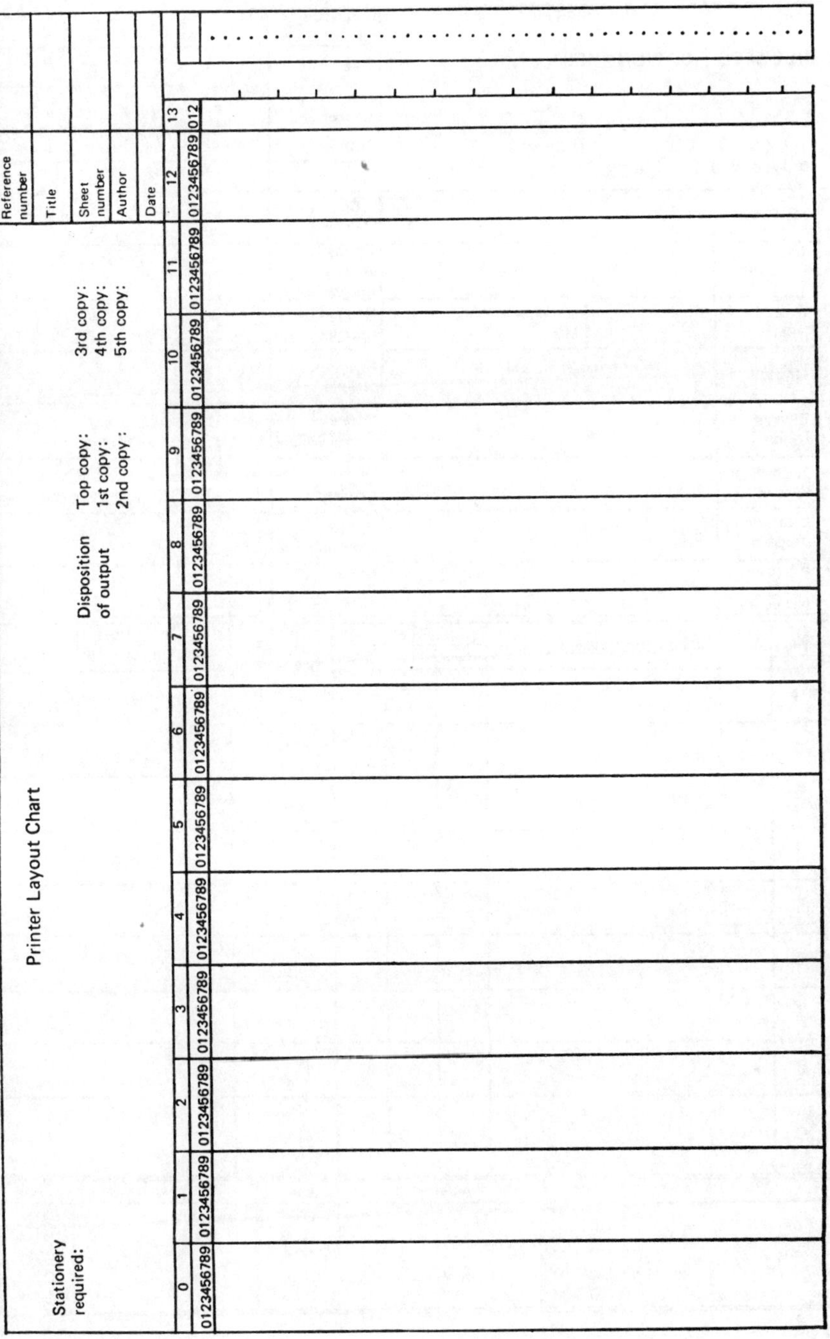

Fig. 16.6. Printer layout chart.

					Reference number		
	FILE SPECIFICATION SHEET				Title		
					Sheet number		
	I = Input; O = Output; T = Transient (or Working); M = Master				Author		
					Date		
General Characteristics	Medium				Estimated file size		
	Structure				Estimated growth/ fluctuation		
	Record length	Fixed	Variable Min. Max.		Number of volumes		
	Units	Word/bytes/characters			Retention period		
	Blocking factor				Number of generations		
	Recording density				Update frequency		
	Creation program				Update program		
	Reorganiz- ation frequency				No. of tracks (mag. tape)	7	9

	No.	Data Element Name	Size			Data Element Reference	Notes (absolute or range of value)
			Fixed	Min	Max		
Record Description	1						
	2						
	3						
	4						
	5						
	6						
	7						
	8						
	9						

	Program name	Use	Program name	Use	Program name	Use
Program use		I/O/T/M		IOTM		IOTM
		I/O/T/M		IOTM		IOTM
		I/O/T/M		IOTM		IOTM
		I/O/T/M		IOTM		IOTM
		I/O/T/M		IOTM		IOTM
Notes						

Fig. 16.7. File specification sheet.

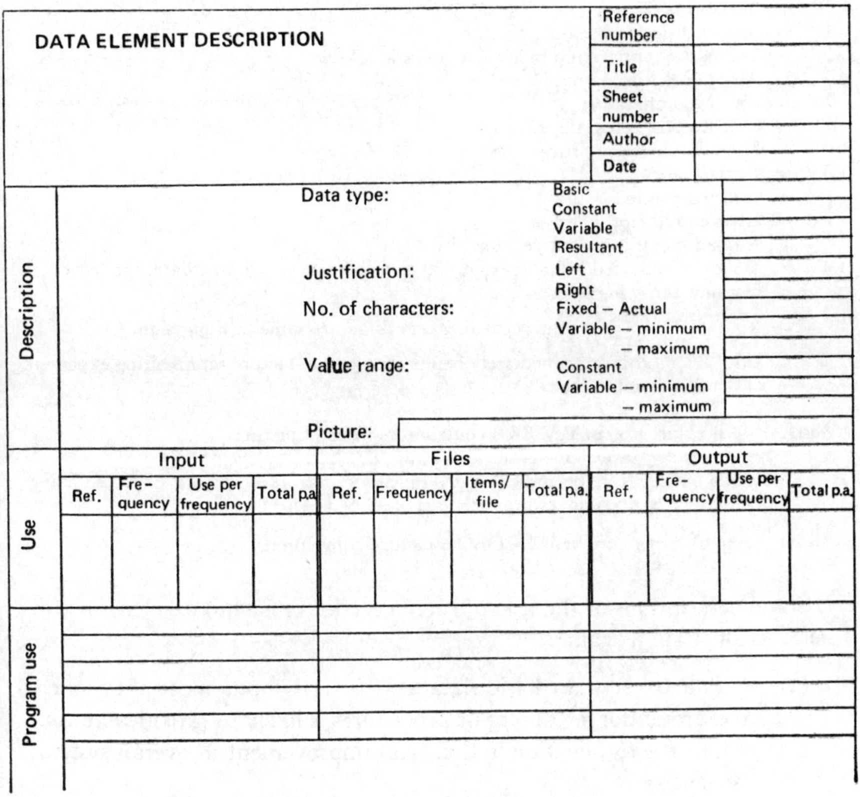

Fig. 16.8. Data element description form.

by a central authority, e.g. the chief programmer or librarian. This name is shown in the 'Title' section of the identification box at the top right-hand corner of the form.

The next sector of the form describes the format of the data, stating whether it is right or left justified, of fixed or variable length and, if the latter, what the minimum and maximum number of characters are. This section of the form also provides a picture of the data element. The picture in the example is completed using the COBOL conventions which are listed in Fig. 16.9. The form then describes the use made of the data element in input, giving both the reference of the appropriate input specification and an estimate of usage. This is followed by similar sections identifying the use of the data element in files and in output.

The final section of the form provides a cross reference to the programs in which it is used.

```
A  = Alphabetic character or space
B  = Space (blank)
P  = Assumed decimal scaling position
S  = Leftmost character indicating an operational size
V  = Assumed decimal point
X  = Any valid character
Z  = 0–9; leading zeros suppressed
9  = Numeric; zeros not suppressed
0  = Inserted zero
,  = Inserted comma
.  = Inserted full stop
*  = Numeric; leading zeros replaced by *
£  = £ sign; if repeated indicates sign will appear once only immediately to left of
     first non zero character
+  = + or − inserted      ⎫
−  = space or − inserted  ⎬ rules regarding repetition are same as for £ sign
                          ⎭
E  = letter E will appear immediately before that part of picture representing exponent
     of floating point value
```

Notes: 1 All values except P,V,E&S contribute to size of picture
        2 A,B,X,9,£, are most useful for picture or character checks
        3 Repetitions can be indicated by repeating picture character or by suffixing
           number of occruances in brackets, i.e. 9999 = 9(4)

Fig. 16.9. COBOL picture conventions.

Some descriptions of the uses of such a record of each data element will serve to illustrate its value.

(1) If data other than basic data are used as input more than once, re-examination of the input procedures is likely to provide elimination of the duplication and also an improvement in overall systems efficiency.

(2) An analysis of the relationships between input/file and output usage of the data element will reveal any misunderstanding or inconsistencies of the system, e.g. if a data element is used as input but there is no record of its use or output, it must be concluded that either it is not needed or alternatively analysis is incomplete.

(3) If different usages of the same basic data call for differing descriptions, a full examination of the basic data is required.

(4) If the usage for a data element is significantly different from the usage of other data elements in the same application, it may indicate the need for a different processing strategy, e.g. reading constant data into core at start of the processing cycle.

(5) The use of data element description is the only way to discern the effect a change in either output or input will have on the total system, e.g. if a source document which is used for the capture of a single data element is to be discontinued, what will be the total impact on the data processing system? This information will become steadily more vital with the trend towards integrated systems.

STANDARDS AND DOCUMENTATION

DATA ELEMENT DESCRIPTION											Reference number	04167
											Title	Part number data definition (PartNo)
											Sheet number	1 of 1
											Author	A. N. Analyst
											Date	3rd January 1971
Description	Data type:					Basic						✓
						Constant Variable						
						Constant						
	Justification:					Left						✓
						Right						
	No. of characters:					Fixed — Actual						7
						Variable — minimum						
						— maximum						
	Value range:					Constant						
						Variable — minimum						A001400
						— maximum						G876500
	Picture:					X 999999						
Use		Input				Files			Output			
	Ref.	Frequency	Use per frequency	Total p.a.	Ref.	Frequency	Items/file	Total p.a.	Ref.	Frequency	Use per frequency	Total p.a.
	STOR 1	Daily	÷100	25,000	0.4 M	Daily	10,000	2,500,000	STOR40	Daily	÷110	27,500
	STOR 9	—"—	÷10	2,500					STOR 41	Week	÷10,000	500,000
	STOR 11	Week	÷1,000	50,000					STOR 42	Mthly	÷10,000	12,000
Program use	ST 41 A											
	ST 46 A											
	ST 48 C											
	ST 11 B											

Fig. 16.10. Completed data element description.

(6) When designing sub-systems for an integrated information system, only the use of data element descriptions will enable costly duplication or omission to be avoided.

(7) In tracing the impact of an error which has entered the system. In this area the use of data element description can be clearly seen and will help prevent additional corruption of the system by enabling a rapid trace of the use of the data to be made.

(8) The use of such a form facilitates exchanges with other data processing users either inside or outside the organisation.

(*i*) *Glossaries.* Particular attention should be given to glossaries. These are of two types, covering data processing usage and company or trade usage respectively. The former can be purchased in pamphlet or book form from a number of sources, the most important of which are listed in the bibliography. Data processing is bedevilled by jargon, much of which

SYSTEM MODIFICATION REQUEST	Reference number	
	Title	
Supplement each section as necessary with additional notes and/or diagrams	Sheet number	
	Author	
	Date	
Name, Position and Dept. of Requestor		
Description of modification required		
Reason for modification		
Estimated cost of modification (one time costs)		
Implications of modification (include effect on running costs)		
Authorization	Requestor: Date:	Systems Dept: Date:

Fig. 16.11. System modification request.

has slightly differing meanings to different people, particularly where they have been working on computers manufactured by different companies. To avoid this problem of internal communication and to assist in its quick assimilation by trainee staff, the terminology of a standard glossary should be adopted.

Data processing staff are in many cases remote from the users of the output from computer systems, who have their own jargon. This problem is particularly acute when the data processing department is staffed with persons who have no previous experience within the company. For the analyst this poses a real communications problem, not only during the analysis of a system, but also throughout the life of a system. Furthermore, if another analyst or programmer is assigned to the maintenance of the system, another learning period will be necessary. To overcome this problem and to help bridge the gap in communications between data processing personnel and the staff of other departments, it is highly desirable to produce a glossary of company and/or trade usage.

(10) *Systems Modification Request* (Fig. 16.11). This form provides a convenient means of formalising requests to the systems department. Its use is fully described in chapter 18.

17
DATA COMMUNICATIONS

The importance of data communication is best illustrated by a comparison of the total time scale of a typical business system which has evolved through the various levels of automation to a modern computer system. Such a time scale (see Fig. 17.1) shows that the very significant improvements in actual processing time are not reflected by the percentage improvement for the total system. In other words the improvement in data processing capability has not been accompanied by much (if any) improvement in the supporting communication system.

It is important to realise that this generalisation is as true for systems which have their initial input and final output phases geographically close to the data processing centre as it is for those systems where the input and output phases are remote from the processing system.

It follows from this that further important time savings in a total system are more likely to arise from the introduction of a data communication system than from further improvement of the data processing phase itself. Thus it is increasingly towards data communications that the systems analyst's attention will be turned. Indeed, some estimates indicate that well over 60% of the computers currently on order will be used either solely or in part for systems involving the use of electronic data communications. Furthermore, in the opinion of many leading experts it is little short of foolhardy to conduct a hardware feasibility without giving consideration to possible future data communication applications.

The communications networks and facilities as well as their economics vary between countries and in some countries between regions. There are, however, common principles underlying all data communications systems and it is these principles which are the foundation of this chapter.

Any data communication system consists essentially of three elements. These are:

- Data collection and transmitting equipment.
- Actual communication facilities, e.g. radio waves, telephone cables, etc.
- Receiving equipment which may be either on-line or off-line to a computer.

In practice, of course, many systems are two-way, i.e. input data is sent

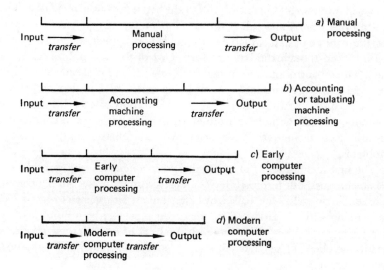

Fig. 17.1. The phases of a typical business system.

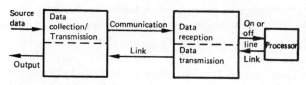

N.B. Data flow may be in either or both directions

Fig. 17.2. Conceptual diagram of data communications system.

to a central computer for processing and output data is returned via the same communications link to the originating point. The underlying concepts are, however, common to both situations. A conceptual diagram of the data communications process is shown in Fig. 17.2.

17.1 Data Collection and Transmitting Equipment

This equipment is in fact a combination of a normal input device, e.g. punched card reader, keyboard, etc., and a terminal device which performs the function of translating the information generated by the input device into a form suitable for transmission. This translation will normally serialise the bit pattern of the characters contained in the input (because of the high cost of providing the line facilities to transmit the multi-bits per character simultaneously over any significant distance). The translation device will also add any necessary control bits to the serial bit pattern created. These control bits (sometimes called the envelope) must be counted in ascertaining the volume of data which can be carried on a given line during a given period, even though they do not form part of the original (or final) message. These control bits are in fact used *inter alia* to synchronize the sending and receiving terminals. A receiving terminal has two problems in receiving data : firstly, it must know the duration of each bit signal and, secondly, it must be able to distinguish the first bit of each character bit group. Two techniques have been evolved for the solution of these problems.

The first technique is the use of the *synchronous* transmission mode. In this technique both terminal devices receive timing pulses from a common source. These pulses are called clocking pulses. In the synchronous mode each message (i.e. a group of characters) is preceded by a number of synchronising characters which establish the basis of reception for all characters in the block. (The block length is a function of machine design and/or programming.)

The second technique is to use the *asynchronous* transmission mode. In this transmission mode single characters are transmitted and timing is re-established for each character, the first bit being a synchronising bit. Because of the need to resynchronise for each character, this mode of transmission is inherently slower than synchronous transmission, especially since a time interval of approximately two bits is also required between characters. It is, however, much simpler (and therefore cheaper) to construct terminals for asynchronous transmission and this cheapness may well offset the savings in line time (and cost) which could be achieved using synchronous transmission.

Asynchronous transmission is particularly suitable for those types of terminal (e.g. keyboards) which work character by character. The codes used by the translation unit of the terminal normally require 5, 6 or 7 bits per character, some of which also include a parity bit for error checking purposes, but codes using from 4 to 12 bits per character have been designed. Thus, using a 5 bit per character code, the actual time needed to transmit 1 character in asynchronous mode will be the equivalent of 8 bits (5 code bits + 1 synchronising bit + 2 interval bits). As stated earlier, it is the actual bit equivalent transmission time which must be used when calculating line capacity or data throughput.

Once data has been translated (encoded) and the envelope added, the next stage in the data communication process is to mount it on to a carrier wave for physical transmission. This process is called modulation and is performed by a *modem* (*m*odulator/*dem*odulator).

Modulation may be amplitude modulation (A M), frequency modulation (F M) or phase modulation (P M). Modulation employing A M and

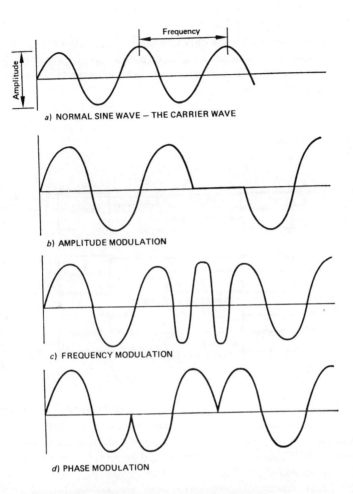

Fig. 17.3. Diagrammatic representation of digital signals imposed on carrier wave using different modulation techniques.

P M is also available. The principles of A M, F M and P M are shown in Fig. 17.3. A detailed treatment of these techniques is outside the scope of the present work and the interested reader is referred to the appropriate works listed in the bibliography. Suffice it to say that, in general, F M and P M are preferred to A M because of the latter system's susceptibility to interference but that the combination A M/P M system is gaining in importance.

17.2 Communications Link

The actual link over which the data transmitted is referred to as a channel. This may consist of one or more wires or a nominated wavelength in the radio frequency range. To the systems analyst four aspects of the channel are important. These are :

Cost.
Error rates.
Mode of operation.
Speed.

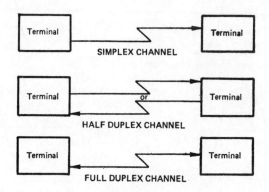

Fig. 17.4. Communications links.

17.2.1 *Cost.* Channel costs will vary according to the nature of the network (e.g. public switched or leased), distance and bandwidth, as well as by country. When considering data communications systems, the analyst must consult the appropriate telephone and telegraph authority to obtain the relevant costs.

17.2.2 *Error rates.* Average error rates are known for all types of channel. Detection of channel errors must be provided and due allowance incorporated in estimated channel costs and timings.

17.2.3 *Mode of operation.* Channels may operate in simplex, half duplex or full duplex mode. A simplex channel only allows transfer of data in one direction; a half duplex channel allows transfer in either direction but not at the same time, whereas full duplex operation allows for the simultaneous transfer of data in both directions. (Fig. 17.4.)

The common belief that half duplex channels are two-wire lines and full duplex channels are four-wire lines is erroneous. The decision as to the number of wires employed in a channel is the responsibility of the authority providing the channel, and unless the terminal device selected presents special requirements, it is no direct concern of the systems analyst.

17.2.4 *Speed.* The speed rating of a channel determines the volume of data of a given coding and envelope structure which can be transferred across the link in a given time and it is thus a vital parameter in data communications systems design. Strictly speaking, it is inaccurate to speak of the speed of the line, as the determinate of speed is the type of modem which can be used with a given channel, and in other than the short term this may be variable. Channels are described conventionally by bandwidth (expressed in cycles or kilocycles per second) and in general the greater the bandwidth the higher the speed at which data may be transferred across the link.

The actual data transfer rate is expressed in BAUDS. In general usage, BAUDS are equivalent to bits per second. The formal definition of a BAUD is, however, *a unit of signalling speed equal to the number of code elements per second.* Therefore, technically, it is only true to say BAUDS = bits per second for codes with equal bit length.

Thus in assessing the actual data transfer rate the line speed in BAUDS is divided by the number of bits per character (code + envelope).

17.3 The Receiving Unit

On arrival at its destination the signal is demodulated. This is the reverse process of modulation, i.e. the stripping of the data from the carrier wave. The receiving unit may be either another terminal (in which case the system is said to be an off-line data communications system) or a central computer (an on-line data communications system).

The remaining technical consideration for a data communication systems analyst is how the mechanics of scheduling the input of data from the remote terminal to the central location are handled. This will depend on :

(1) The channel network.
(2) Whether the system is on-line or off-line.
(3) Whether data transmission is by batch or on an 'as required basis'.

The major types of network are shown in Fig. 17.5.

With off-line systems it is usual for transmission to be in batches and pre-arranged schedules can then be provided to the operators at each terminal. Furthermore, voice communications are normally established before data transmission commences and thus there is little problem in maintaining control over the input to the system.

With an on-line system the complexity of the problem will depend on whether input is batched or on an 'as required' basis. For batched input, regular schedules can be developed and, if necessary, off-line connections established before data is transmitted. With 'as required' use, one technique is for the central computer to invite each terminal in turn (or on an

a) One to one (simplex, half or full duplex)

b) Centre to terminals (each line is simplex)

c) Terminals to centre (each line is simplex)

d) Terminals to and from centre (each line is half or full duplex)

N.B. Direct connections may also exist between terminal locations

Fig. 17.5. Basic data processing communications networks.

eccentric basis according to estimated volume of traffic) whether or not it has data to transmit. This process is known as polling. If a terminal has no data to transmit the next terminal is invited and so on. If the terminal is ready to transmit, a connection is established and the data transmitted. An alternative technique is for a terminal which has data to transmit to generate an interrupt signal in the central processor.

17.4 Data Communications and Systems Design

In designing a system which utilises data communications, all the principles outlined in Part I of this book will be applicable. In addition, however, there are a number of additional considerations which must be taken into account in a data communications system. It is only these *additional* considerations which are considered in this section. These additional considerations may be identified as location (actual and required), volume, response time, coding system and accuracy of the date.

17.4.1 *Location* (actual and required)

The first stage in designing a data communication-based system is to ascertain the location of the source data, the processing capability and the point or points at which the processed data is used. This is the first step in determining which of the types of network identified in Fig. 17.5 will be adopted. If this analysis reveals, for example, that source data and final output are required at the same location, but that it is necessary for processing to be carried out elsewhere, particularly close attention should be paid to the economic viability of the system.

17.4.2 *Volume*

The next stage is to estimate carefully the volume of data which the system is to be designed to handle.

It is first necessary to estimate the number of messages per day and here it is advisable to count the numbers of different types of message separately, since the average length of each type is likely to be significantly different. Having ascertained the average number of messages of each type per day and noting any daily, weekly, monthly or seasonal distribution patterns, the next process is to estimate the average number of characters per message. Like the estimation of the average number of messages, statistical sampling techniques-may be used for this calculation, but the analyst is warned of the dangers of using a sample too small to produce a valid result. In any case of doubt, the advice of a qualified statistician should be sought or reference made to one of the standard statistical works available in reference libraries, which show the calculation of minimum sample sizes needed to produce a stated degree of accuracy from a random sample. In calculating the number of characters it must be remembered that embedded spaces, carriage control characters and other non-printed information will need to be transmitted and must therefore be included in the message length as well as the actual data characters.

With these statistics of number of messages and average length of message, the analyst is almost ready to ascertain the volume of traffic. Before doing so, however, due allowance must be made for re-sending data found to be in error and for transmission errors. This latter point brings the analyst to the first of many 'chicken and egg' situations. To calculate volume of traffic and thus decide on the type of line required he must know the line error rate which in turn is specific to a type of line. Furthermore, it assumes a particular code structure! It is therefore necessary to go through a reiterative process of calculation and examination of a number of feasible solutions. Many works on the subject advise the analyst to calculate on the basis of average loadings on regular busy days. The analyst is advised, however, to check his calculation against a 'most busy day' situation and use queueing theory to ascertain what delays this load will generate. These maximum delays (and the statistical probability with which they will occur) should be presented to management for a decision as to whether they are acceptable. A further parameter which must be considered is that of expansion. When planning an on-line data communications system, a time scale of at least five years should be considered because of the nature and volume of the expenditure involved.

17.4.3 *Response Time*

The response time of a system is the maximum period after the origination of the data for which the recipient is prepared to wait for before requiring it. It is easy to say that this is 'as long as it takes at present', but this may ignore sound commercial reasons for requiring the data sooner. Suffice it to say that this point will require the analyst to exercise tact and judgement in ascertaining the validity of claims for required response times.

Having ascertained source and destination of data, the volume of traffic (allowing for expansion) and the maximum response time, the analyst is in a position to calculate the feasible alternatives for meeting these requirements.

17.4.4 *Coding Systems*

As mentioned above, codes utilising from 4 to 12 bits per character are in use. From the analyst's point of view the design parameters are that the system must if possible accept data in the form it arises (e.g. if it arises as a by-product of a punched card operation the communication system must use punched cards codes as input) and present it in the form it is required. Provided the code used provides a sufficient character set and line capacity is adequate the analyst has no other practical concern.

17.4.5 *Accuracy of Data*

The question of line errors and indeed input errors has already been considered in conjunction with the establishment of data volumes, but before finalising the design the analyst must consider what the residual tolerable level of errors in the system will be and what implications any alteration in this level will have on network design.

18
SYSTEMS MAINTENANCE

Most systems become obsolete the day they become operational. That is not to say they become without value to the organisation, but almost invariably changes are necessary to the new system right from the start. The systems analyst therefore rarely sees a system 'completed' but is involved almost continuously in maintaining the systems he has implemented at a satisfactory operational level.

Why are such changes necessary? Some of the main reasons are:
- Oversight on the part of the systems analyst at the design stage.
- Misunderstanding by the systems analyst of the requirements of the user department.
- Insufficiently tested systems before going operational.
- Change in the user departments' procedures.
- Desire of the user department to experiment.
- Company policy change, e.g. pricing structures, credit policies, buying policies, etc.
- Change in legal requirements.

Many changes are made unnecessarily or without thought of the implications, such as cost of the change, and clearly these should be minimised. But many cannot be avoided, such as a new income tax structure. So inevitably systems maintenance forms part of the permanent workload of a systems department. After every new system is installed an allowance has immediately to be made for future maintenance of the system.

Clearly, after a number of systems are operational, a significant proportion (we have met cases of 60%) of the total workload is being allocated to system maintenance. One soon comes, therefore, to the conclusion that there is a finite number of systems which can be installed with a given level of manpower. This is confirmed by observation of actual computer installations. The first few years of most companies' computer systems are marked by the creation of a host of new systems by a relatively (compared to later) small team. After a few years, however, progress seems to slow down as more and more time is needed to keep the old systems going. Eventually, it becomes more sensible to rewrite the systems completely rather than continue amending them. At this point it becomes clear that only with the addition of more staff can the old systems be rewritten and truly new systems be created, so an enlargement in the staffing level takes place. Once this policy decision has been made (that some new systems should be developed as well as maintaining old systems), the trend is set. A steady expansion in staff numbers becomes the pattern.

This has other cost implications, too. Not only is extra space needed, but more testing time on the computer, which usually has to be enlarged to meet the needs of the increased staff, and so on. We can see, therefore, that systems maintenance is an important contributing factor to the characteristic pattern of rising costs of almost all computer installations. Anything that can be done, therefore, to reduce it will benefit the whole organization.

How can we minimise the amount of systems maintenance? There are a number of approaches which can be taken, all of which can contribute to achieving this aim :

18.1 User Involvement

It is essential for successful systems analysis and design that the person(s) who will eventually be making the use of the system be actively involved in its design. This should go without saying, but it is surprising how many systems are installed only for it to become soon apparent that the user was not really *involved*. This lack of involvement manifests itself in many ways, as discussed in the special section on this topic in chapter 19, but there is nothing more detrimental to the overall efficiency of the computer installation than the high amount of systems maintenance which invariably ensues. One must work at the detailed level with user departments, not only during analysis but also at the design stage.

18.2 Modularity of Systems

Some maintenance is inevitable. It is therefore wise to be as well prepared as possible when it does come so that the task itself is accomplished with the least trouble. A major step towards this aim can be achieved by the 'modular approach'. Systems should be constructed in logical units of a limited size. Examples of such units are input/output routines, error cycling, report printing, etc. Further examples, plus guidelines for constructing modules, are given in Part I, 5. The main point which we wish to emphasise here is that maintenance is much easier if the system has been constructed in modular units. The effects of a change are more easily discerned, and limited, by the use of modules. Furthermore, a module into which a change has been introduced can be tested separately. Modifications are easily documented and are understood by others, who may at a later stage have to examine the system unaided.

The modular approach is simple in concept, but has immense implications of real value and should therefore be pursued.

18.3 Modification Procedure

It is important to have a control procedure to prevent unnecessary changes being made. Most user departments have no idea of the implications of asking for even a small change, such as a change in the printing format of a report (this would consume time of the systems analyst, programmer, operators and computer). Often this desire for modification is made out of enthusiasm and there are occasions when one is willing to pay some cost just to keep a user happy. But of course even in this sort of situation one must retain control; otherwise, by accepting all suggested modifications, one will be faced with an ever-increasing maintenance task.

The control procedure should be based on the following key points:

(1) Any request for modification must be put in writing and signed by a responsible person (section-leader or above) in the user department.

(2) The request should be evaluated by the systems analyst in terms of cost for doing the job and in terms of effects of the change on other systems, computer time, etc.

(3) The cost figure should be presented to the user department and their agreement obtained that the modification is worthwhile. If the modification shows direct savings, then there is no problem. If the benefits are more vague, then it is essential to obtain agreement in writing (i.e. a signed acknowledgement). Putting a signature to a cost figure has an amazing psychological effect—people think much more carefully about the justification for a sum, if they know they can be identified later as the originator of the expense.

These three key points can conveniently be moulded into one simple procedure by having a special document, which includes the following information:

- The name of the person requesting the modification.
- A description of the modification required.
- The reason for the modification (including estimated benefits).
- The estimated cost of making the modification.
- A statement of the implications of the modification, e.g. increased running costs, effect on other systems, etc.

A suitable form for this purpose which meets the standards for documentation developed in this book is shown in Part III, 16. Such a procedure can have a significant beneficial effect in minimising the systems maintenance load.

18.4 Organisation

At the beginning of this section we discussed the legacy which remains with every systems analyst after he has implemented a new system. He is constantly being interrupted in his other work to make modifications to his operational systems. This is a very unhealthy situation for the company and the analyst for several reasons. Firstly, it is frustrating for the analyst

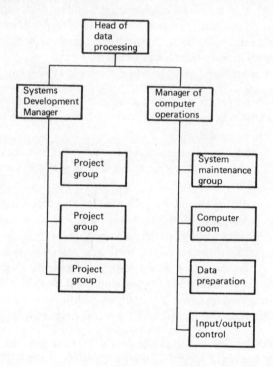

Fig. 18.1. Typical organisation structure of a data processing department which has separated systems development and maintenance functions.

to be interrupted from time to time in his development work on new projects. Secondly, it gives scheduling problems. It is difficult to anticipate maintenance work and to incorporate it into an analyst's work schedule. Thirdly, it encourages dependence on the individual. Modifications tend to be made in a hurry and are often poorly documented. So a system which has been running for some time usually is not truly represented by its documentation; only the analyst who made the changes can thoroughly understand the way the system works. If he leaves the company, therefore, severe problems are likely to be encountered when further modifications become necessary. This can be counteracted by strict adherence to a standards policy, as discussed in Part III, 16. There is, however, another way which can largely eliminate this problem, namely the setting up of a special organisational unit to carry out all system and program maintenance.

The principle behind this approach is to separate development work from maintenance work. The maintenance section then becomes fully responsible for all modifications to operational systems. The benefits of this approach are substantial. Firstly, the development personnel are not disrupted from their work. Secondly, job scheduling becomes easier. Thirdly, systems are in a much better condition to go operational when they are 'finished'. This is because there is a psychological incentive to remove errors from a system you have developed if you are going to hand it over to someone else. It is a matter of personal pride in your work. Fourthly, the standard of documentation is higher, because the maintenance group have the right to refuse responsibility for a system if they feel it is not sufficiently well documented to maintain it. This implies an 'acceptance test' which is a standard control in any other field and should certainly be used for computer systems.

The structure of a typical data processing department adopting the above principle will look something like Fig. 18.1.

There are, of course, a variety of ways in which these functions can be organised. The main point of the above diagram is to show that the System Maintenance Group reports to the head of Computer Operations, the 'production' side. The division between development and maintenance is then complete in order to guarantee the benefits listed above.

19
USER INVOLVEMENT

If an experienced systems analyst is asked to name the biggest single factor which determines the success or failure of a system, he usually answers 'User involvement'. He is referring to the desirability of having the user of the future system work closely with the analyst right from the start of the project. This has many benefits, some of which are:

- It ensures that the analysis and design do correspond to the requirements of the user area.
- It ensures acceptance of the system when it becomes operational. If the user himself has helped create the design, he cannot resist its use afterwards.
- It provides built-in training capability for the user department. Although there will probably be aspects of the system best explained by the systems analyst, the user representative(s) can do much of the system introduction.
- It provides automatic follow-up. It is very valuable to have a member of the new department who thoroughly understands the system 'on site' the whole time.

There are many other benefits, such as good public relations for the systems department, training in user areas for systems analysts, etc. The question therefore arises: 'How does one "involve users"?' What exactly do we mean by 'involvement'? The answers to these questions depend on the level of responsibility of the user. If we are thinking of top management whose company divisions are being affected by systems activity, then these are involved in a different way from the payroll clerk whose job is about to be assisted by the computer.

Let us therefore identify the following categories of user:

- Top management.
- Line management.
- Non-management personnel.

These are involved to differing degrees and in different ways:

19.1 Top Management

The involvement of top management occurs mainly in the following areas of responsibility :

(1) Appointment of systems management.
(2) Approval of organisational structure within systems department.
(3) Setting of objectives for systems department.
(4) Allocation of resources and approval of budgets.
(5) Monitoring of progress by review at major checkpoints.
(6) Open support for work of systems department.

These responsibilities are high-level management functions, which apply to any company department. However, the responsibility described under (6) above is rather different from other areas. The purpose here is to give the systems department the prestige necessary to carry out a project which is in the company's interest, but which may not be appreciated by individual departments. Of course, it is part of the systems analyst's job to convince people of the value of systems improvements, but he cannot always be successful in this task, because personal motives of company staff are involved as well. In such cases, knowledge that the project is supported by top management will greatly increase the chances of acceptance of the system.

19.2 Line Management

Line management are often the people most affected by new systems. For this reason it is essential for success of the system that they support the project. This support is best achieved by involving them in the following ways :

(1) Give them an appreciation of the goals of the systems department. This can be achieved by personal discussion and by periodic formal presentations to inform them of and to involve them in systems development plans.
(2) Keep them informed of progress. It is a good idea to be quite frank about the progress of systems both those under development and those which are operational. This will prevent line management expecting too much too soon, and will often gain valuable additional support if they see a way in which they can help.
(3) Ask them to appoint liaison personnel and/or project team members. For every systems project, no matter how small, it is highly desirable that the department(s) affected by the project appoint some person(s) to be the main contact man. This person's function

will be to ensure that the requirements of his department are being met, and to assist the project team in gathering data and implementing the system. For medium and large projects, it is desirable that this person becomes a full-time member of the project team.

19.3 Non-management Personnel

The important means of involving non-management personnel are:

(1) Give them an appreciation of the goals of the systems department.
(2) Keep them informed of progress.
(3) Consult them constantly during the fact-gathering and design processes.
(4) Train them in the use of the system.
(5) Stay in touch with them constantly during the running-in of the system, if possible working with them.
(6) Consult them after implementation as part of the post-evaluation phase.

It is not sufficient to pay lip-service to the idea of user involvement; active steps must be taken along the lines described above.

PART IV
Project Control

20
PROBLEMS OF IMPLEMENTATION

The term 'implementation' in data processing is used to cover a variety of different meanings, ranging on the one hand from the simple conversion of an existing computer application to a revised or extended application, to the complete changeover from one type of hardware to another, on the other hand. In practice the problems to be overcome tend to be much the same and the differences are of degree rather than principle. By 'implementation' the authors of this book mean the process of converting the system design into an operational system. This chapter outlines some of the major problems which arise in this important phase of systems work, i.e. the generalised case of 'getting the system to work', and describes measures which may be used to minimise these problem areas.

Firstly, however, let us consider some examples of implementation and the types of problems which arise :

(1) *Implementation of a computer system to fulfil a completely new requirement*, i.e. the organisation has previously had no system (manual or otherwise) having this function.

In many ways this is the easiest type of implementation for the analyst. There is in general no problem of reconciliation with an existing system or of staff being concerned with the continued operation of the existing system whilst the new system is being 'run in'. On the other hand, systems of this type are frequently needed by a fixed date, e.g. a new financial year or the introduction of decimal currency, and as there is no back-up system available, this requirement can be a rigid taskmaster.

(2) *Implementation of a computer system to replace a manual system*

The major problems normally encountered in this type of conversion are file creation and training of user personnel. File creation poses two problems. Firstly, there is the creation of an accurate data file with its consequent high demand on data preparation, verification and the need to check manually print-outs of the file after its preparation. Secondly, there is the difficulty of maintaining such files in phase with the manual system files during parallel running and/or the cut-over period to permit reconciliation. Personnel training will vary according to circumstances, but in this situation may well involve people who not only have never been involved with computer systems before but may in addition have a conscious or unconscious fear of 'automation' and possibly also feel that their job is endangered.

(3) *Implementation of a computer application system to replace an existing computer application system using the same hardware*
The ease with which this type of implementation is accomplished will usually depend on whether or not the same data files are to be used. If the files are the same, the problems are not too severe. At least some of the user staff involved will be familiar with the requirements of the new system and, unless the system being replaced has a particularly unfortunate history, will generally be a useful reserve of informed support.

(4) *Implementation of a change in systems software*
In medium-size and larger installations, this type of implementation will normally be carried out by a system software specialist. The systems analyst must, however, be acquainted with the scope of the problem and its possible effects on applications. These effects may influence systems timing, e.g. when off-line printing is introduced, or file organisation possibilities, e.g. when a new or revised operating system is generated.

(5) *Conversion from one computer to another*
In many ways this is the most difficult type of implementation problem of all. It took many companies over three years to recover from the problems encountered in changing from second- to third-generation hardware and even at the end of that time many organisations were still using emulation or simulation to run second-generation programs. Yet, properly planned, there is no need for the problem to reach such horrifying proportions. Planning for conversion should, indeed must, start with the original system design—it is at this stage that all the principles of flexibility should be built into *every* system and be supported by good standards and documentation.

The second major step in reducing the problem to manageable proportions is the careful selection of the replacement hardware. The cost of the conversion process itself must be taken into account when assessing the economic viability of the change and due regard should be given to the existence of emulators, simulators or translators for existing hardware and programs.

From the above brief analysis of some of the types of implementation process with which the analyst may be involved it can be seen that personnel training and file conversion are the two areas in which problems are most likely to occur. A further problem area is, of course, programming. The remainder of this chapter examines each of these problem areas.

20.1 Personnel Training

With the introduction of any new system it will be necessary to train or retrain members of the staff of all departments concerned to operate the system. This area is, however, often neglected and this neglect can lead to a total collapse of the system.

The first point to remember is that *all staff concerned* should be trained. This will clearly involve staff in the user department, but also staff in any other affected departments must be trained in the requirements of the new system. The same is also true of staff within the data processing department who will be responsible for the operation of the system. This training will, in fact, cover two separate though related aspects, i.e. general information about the aims and objectives of the system and the detailed training in the procedures involved in the operation of the new system.

The first aspect of training in the general aims and objectives of the new system is a vital aspect of successful implementation and the presentation made for this purpose should be to as wide an audience as possible. This phase of training, or rather education, should stress the overall aspects of the new system from the total company point of view and illustrate the tangible and intangible benefits that will result from its introduction. Visual aids, including flipchart boards and overhead projectors, are ideal for presenting this type of information in a way which will both interest and educate the watcher. Time spent in preparing and giving such presentations will be more than compensated for by the saving in time during the critical phases of implementation due to the improved understanding and motivation of all concerned.

The second phase, the detailed training in the operation of the system must be directed at those persons who will actually do the work, but at the same time ensure that all levels of supervision concerned have the detailed knowledge necessary to exercise their supervisory responsibilities. This phase of training must be supported by good user-oriented documentation (see page 34) which should be prepared by the analyst with the active assistance of the user department. Full opportunity must also be given for operating staff to use the input documentation, coding systems, etc., involved in the system *before* either parallel running or live running begins. This means that in planning the implementation of a system the analyst must ensure that adequate supplies of all necessary documents are available for this phase of the training as well as for parallel and live running. This process of familiarisation training will normally take place at the same time as systems testing, and this stage of a project, therefore, represents a particularly heavy workload for the analyst or analysts concerned.

The training of data processing staff must also not be overlooked and appropriate action should be taken to ensure that data preparation, input and output control staff, as well as computer and ancilliary machine staff, are all fully aware of the actions they must take and the time constraints applicable before parallel running begins. Particularly valuable results can be achieved by arranging informal tours of the computer department by user staff and of the user department by computer operations staff. This promotes an understanding of each other's real and potential problems and enables good working relationships to be established which will prove invaluable when operational problems inevitably occur.

One of the problems mentioned above is that user staff may feel their job security is endangered by the new system. There is no easy answer to this problem. No one will be reassured by bland promises, and unless the organisation is large enough to operate a 'no redundancy' policy this can present the analyst with an acute problem. In any systems investigation in which there is likely to be some effect on staffing levels or a significant effect on job content (particularly if staff who are members of a union are concerned) the analyst should seek the advice of the personnel department and the appropriate manager at the earliest possible opportunity.

Another aspect of staff training which will affect the analyst during the implementation of a system is that he must ensure that the data processing staff are fully conversant with both the hardware and software involved. This is particularly important when hardware or an application package is to be used for the first time. It is not sufficient for the analyst to assume the proficiency—he must check and if necessary initiate action to correct any deficiency.

20.2 File Creation and Conversion

The main problems involved in creating a computer file to replace a manual file have already been identified, namely the volume of work involved in the physical creation of the file and the difficulty in maintaining the computer file in phase with the manual file during parallel running. It is the experience of the authors that this is one of the most frequently neglected areas of implementation planning. This is unfortunate, as a failure to pay proper attention to this problem may well cause much of the value of a parallel run to be lost because of the impossibility of reconciling manual and computer procedures. For this reason alone it is a prudent move to assign adequate resources to this problem both in terms of quantity and quality. Ideally, those clerks most familiar with the manual data files should be given the task of vetting the computer files and maintaining the relationships between the old and new files during the changeover period.

With regard to the problem of the volume of data preparation which is involved in implementing a system involving new data files, few data preparation sections are (or should be) staffed and equipped to handle such peak loads. A method frequently adopted is to attempt to schedule this peak load on to the section, but this approach is almost certainly doomed to failure, since the supervisor is naturally predominantly concerned in meeting routine operational deadlines. Consequently, the preparation of new data files tends to take a relatively low priority, with a consequent adverse effect on the new system target date. Nor is it usually wise to rely on overtime or excess hours' working to meet such a peak load, since most data preparation staff are reluctant to undertake extra working hours. The only practical solutions are either to staff the data preparation section to cater for peak loads (which may be feasible during periods of rapid expansion) or to have the work done by an outside agency. If this latter solution is anticipated, the analyst must include adequate provision for the charges involved when preparing his cost estimates for the system.

20.3 Programming

Programming policy will, of course, normally be the province of the chief programmer, but it impinges so heavily on the implementation of a project that some major points are mentioned here.

The programming language used and the concept of modular programming are matters of vital concern to the analyst as well as to the chief programmer. In addition to safeguarding the ability to maintain the programs (in conjunction with good documentation), they also provide the capability for rapid alteration during systems testing and parallel running. For this reason, the systems analyst should work in close co-operation with the programming staff who are connected with the project. He will be required to clarify points in his program specifications and should monitor progress to assess the effect of programming problems on the project completion time.

These then are the major problem areas which are likely to be encountered in system implementation and steps which may be taken to minimise them. The real solution to the problems of implementation is, however, forethought. A problem anticipated is a problem half solved. It is prudent, therefore, for the analyst to devote considerable time to the identification of potential problem areas by discussions with both user and data processing staff and allocating sufficient resources to overcome them. The other key point in successful implementation is the existence of a suitable planning and monitoring technique to enable the analyst to control the project effectively. Suitable techniques for this purpose are discussed in the following chapter.

21
PLANNING FOR IMPLEMENTATION

The task of a systems designer does not end when the new system, whether it be manual or computer oriented, is specified and accepted. It is his responsibility to turn that design into a working system. It is, however, in many cases a failure to understand the factors involved or the ineffective management of resources available which leads to many projects falling behind schedule during this critical phase.

Both these problems, which are common to the implementation of any project, can be alleviated by the use of a sound plan of action developed in the form of a good planning technique. The first stage in the application of any such technique will be identification of all the *activities* which collectively form the implementation of the project. This analysis of the tasks to be undertaken completed in narrative form is in itself an aid, which will assist in the satisfactory completion of the project. The next stage in the process is to make an as accurate as possible estimate of the time needed for the completion of each activity.

Having completed this preparatory work the designer can either proceed to prepare a plan in narrative form or use one of the planning techniques that have been developed to present the plan in a visual form.

The simplest of these techniques is the familiar Bar or GANTT chart (see Fig. 21.1) which consists of a horizontal bar to represent each activity in the project. The length of each bar is proportional to the time that it will take and the horizontal axis of such a chart serves as a time scale. Each activity is therefore shown at the time in the project at which it is planned to be worked upon. If unshaded bars are used for activities and these are shaded when completed (either in full or in measurable part) then the use of a vertical cursor will quickly show which activities are behind schedule and which are ahead of schedule. The major advantages of GANTT charts are their simplicity and that they show the activities on a time-related base. The major disadvantage of this technique is that it does not show the interdependence of the various activities. Thus, although a visual inspection will quickly reveal the current status of each activity, it will offer no guide as to what effect a delay on one activity will have on the total project.

It was to overcome this fundamental disadvantage that the network diagram techniques were developed. The generic name (network) for these techniques is used here in preference to the names PERT (Program

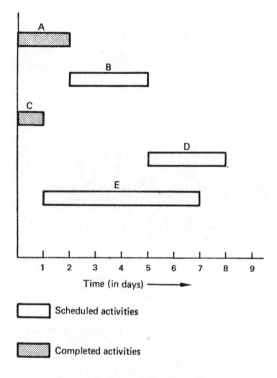

Fig. 21.1. A simple GANTT chart.

Evaluation and Review Technique), CPM (Critical Path Method), MPM (Metra Potential Method) Precedence charts, Arrow diagrams, etc., for, whilst many of these have particular features, all are based on the same general principle of using a diagrammatic representation to show the interrelationship between activities. A simple network diagram (Fig. 21.2) illustrates this point. In this case activities are represented by the arrows joining the nodes and the nodes themselves are used to represent events. Events are the start or completion of a particular action, e.g. *computer delivered*, whereas activities are the elements of the project, e.g. *write program A*. Note that the convention of using nodes to represent events and arrows to represent activities is reversed in some of the network techniques (notably MPM).

Network diagrams overcome the major disadvantages of GANTT charts in that they do reveal the interdependence of activities, e.g. Fig. 21.2 shows that event 4 (the start of activity G) cannot take place until activities F and B have been completed. They also have a major advantage in that adding the times of the activities on each route or path between the start and finish of the project will quickly reveal the *critical path* for the project. This is the series of activities in which any delay will cause a delay in the completion of the project unless corrective action can be taken. It is also possible to calculate the *float* time for each activity not on the critical path, i.e. that amount of delay to which any activity can be subject without any effect on the final completion of the project. For large-scale projects these calculations can be performed by computer using the programs provided by the manufacturer or a specialist software house.

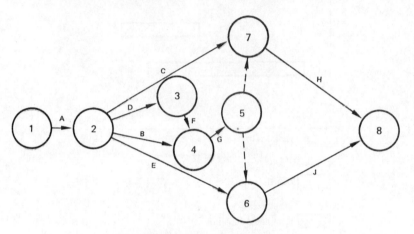

Fig. 21.2. A simple network diagram.

Although network diagrams do possess these advantages, they also have serious limitations. Firstly, the current status of the project cannot be shown on the network diagram. Secondly, there is a need to include dummy activities to show inter-relationships that do not involve work or effort; two dummy activities are shown in Fig. 21.2 linking event 5 to event 6 and event 5 to event 7 respectively. Thirdly, the manual preparation of a network is a tiresome chore, especially since nearly all alterations will require the whole chart to be reconstructed. Fourthly, a complex network is generally not acceptable as a report for line managers.

Network diagrams, although simple in theory, are not, therefore, widely used except by staff regularly involved in the planning of major projects, e.g. in the civil engineering and shipbuilding industries. For systems designers the chore of preparation and updating, together with their lack of acceptability as a management reporting document, has limited the acceptance of the technique, especially since in overcoming the inherent limitations of the GANTT chart the network diagram has lost the principle advantage of the simpler system in that it relates activities to a time scale.

A full treatment of network diagramming techniques is given in the

SCHEDULE OF ACTIVITIES

System: New Computer System

reference	activities description	estimated time (days)
A	obtain agreement to proposed system	1
B	prepare and agree training programme	1
C	train user staff	10
D	write program	6
E	test and debug program	6
F	parallel run	3
G	system acceptance tests	1
H	order and await new stationery	7

Prepared by   Date

Fig. 21.3. Schedule of activities and estimated times.

appropriate works included in the bibliography. To acquire the advantages of the network diagrams whilst retaining the benefits of the GANTT charts a further technique has been developed. This technique has a number of titles, including Job Progress Charting and GASP (Graphical Procedure for Analytical and Synthetical Evaluation and Review of Construction Programs). The convenience of the acronym has led us to adopt it in this chapter.

Basically a GASP chart may be regarded as a GANTT chart which has been extended to show the relationships between activities. A simplified example of a computer system implementation will show how a GASP chart is constructed and illustrate the advantages of the technique. Like all other planning techniques the first task is to prepare a schedule of activities and estimated times for their completion (Fig. 21.3). Stage 2 is to identify the inter-relationships between the activities on the schedule. Thus in the example no other activity can commence until activity A is completed. Similarly, activity C will depend on the completion of B, activity E on D, activity F on H, E and C and activity G on F. A chart can now be prepared showing horizontal bars for each activity, the length of which is proportional to the estimated time (Fig. 21.4). The relationships between the activities are then added to the chart (Fig. 21.5) using solid vertical lines to indicate where the end of one activity represents the commencement of one or more subsequent events, e.g. the line joining activity A to activities B, D and H in Fig. 21.5. Where there is a slack time between the completion of one activity and the commencement of the next related activity (because this subsequent activity depends on another activity), the first activity is extended by dotted lines and linked to the subsequent activity by a vertical dotted line. This is shown with activities C and H in Fig. 21.5, both of which have to be completed before the commencement of activity F, which cannot be started until activity E is completed.

Thus by adding the relationships to the chart all activities for which a time float exists are immediately identified. The path along which there is no time float (A, D, E, F, G in Fig. 21.5) is, of course, the critical path.

The GASP technique does, therefore, retain all the advantages of the GANTT chart (principally simplicity, its ability to show easily the current

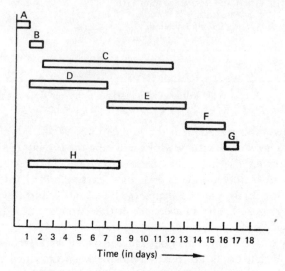

Fig. 21.4. Partially completed GASP chart.

status of the project and the use of a time scale) and also provides the major advantages of a network diagram in identifying the critical path and the relationships between activities. Tests we have carried out in Germany and Italy indicate that the learning period for the GASP technique is considerably less than that for the network diagram. It is clear that the charts presented will be more readily understood and indeed the basic similarity to the well-established GANTT chart will ensure its acceptability for reporting purposes at all levels of management.

The GASP technique is an ideal method for a systems designer to use in planning the implementation of a project. It meets the essential requirement as a planning aid without imposing a heavy demand on time for its preparation. Moreover, it saves time by being an ideal reporting technique.

Furthermore, use of a plastic strip charting media or magnetic strips on a metal chart board to present the GASP chart does provide it with the flexibility to be used for resource scheduling as the individual activities can

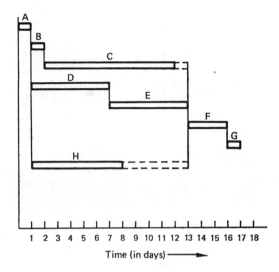

Fig. 21.5. Completed GASP chart.

be drawn to suit the task of one individual and these strips adjusted to the manpower available. The judicious use of colour in preparing the chart can also be an advantage, as separate colours can be used to separate 'waiting' activities, e.g. await delivery of new stationery, from 'active' activities, e.g. writing a program. This can focus attention on those areas where the application of additional effort in the form of extra staff or material can result in overall time savings.

The delays which occur during the implementation of systems (and particularly computer systems) can in many cases be eliminated by sensible planning. The use of a suitable technique greatly facilitates this process and the simplicity and effectiveness of the GASP technique make it ideal for this purpose. No organisation can afford to waste either human or material resources and the implementation of systems on target plays a major part in ensuring their effective use.

22
MANAGEMENT CONTROL

The previous discussions in this section on project control have concentrated on the role of the systems analyst in planning and implementing project activities. At a higher level, of course, the systems analyst himself is controlled as part of the overall systems effort, which in turn is monitored by top management in the light of company goals.

Clearly a detailed discussion of such management control is beyond the scope of this book, as it would take us into realms of the theory and practice of general management principles. What we would like to outline here are those aspects of good management which have particular bearing on the control of systems development.

Firstly, let us examine the principles which should determine the way in which systems development is controlled. Then we shall be in a position to establish a suitable control mechanism. The main points to watch are:

- Is there a link to top management?
- Is provision made for ensuring that systems development really contributes to the goals of the company?
- Is there a means for involving general management in systems development?
- Are there regular checkpoints at which top management can monitor progress?
- Does systems management have the means to co-ordinate the different projects under way?
- Does systems management make effective use of specialists who are not attached to any one area of development?

After observing many different attempts to meet these requirements, it is the authors' opinion that the best approach to follow is the 'Steering Committee' method. This approach is based on a group of fairly senior management who meet periodically to set goals and review progress of systems development. The person responsible for systems is a member of this committee, the chairman of which is a member of top management. So the make-up of the committee is something like:

- Member of top management (chairman).
- Department manager A.

- Department manager B.
- Department manager C.
- Chief auditor.
- Head of systems department.

This Steering Committee would meet about once a month for half a day and would concern itself with resolving problems of systems development, such as conflicts of priority, lack of resources, etc. It would itself report at appropriate checkpoints to top management, who would be involved in the way described in Part III, 19 on user involvement. Suitable checkpoints could be events such as placing an order for a new computer system, the completion of a phase of systems development, a consultant's report, year-end results, etc. In this way a structure is provided which ensures top management has an appropriate control mechanism for controlling the planning and development of all systems work. Overall directives will be given to them by top management and by the Steering Committee, but the implementation of these policies is their responsibility.

Apart from the usual management capabilities, there are some special features of managing systems development :

(1) *Multi-project environment*
The problem of planning and controlling many projects simultaneously is characteristic of systems development. This implies problems in the areas of :
- job scheduling
- resource allocation
- peak loads at non-regular intervals

These characteristics suggest a strong similarity to the management of large construction projects. Perhaps the similarity is not so surprising, if one views systems development as the gradual building up of a structure for company activities to work within.

(2) *No two projects are alike*
Although the task of systems analysis and programming should follow a standard methodology, the problems to which they are applied are new each time.

(3) *Systems development becomes progressively more difficult*
This factor is very common. When a company first introduces computer systems they are usually fairly simple and easily implemented. But as time goes on and more systems are created, a higher degree of dependence between systems creeps in. This makes it more and more difficult to introduce a new system.

(4) Performance is difficult to assess

It is notoriously difficult to judge the effectiveness of a systems analyst. Apart from the problem that he will probably be working on several different projects within the evaluation period (one year perhaps), the work itself is difficult to assess both from a qualitative and a quantitative point of view. Many attempts have been made to identify the factors determining the performance of a systems analyst. Most lists include:

- The success with which systems implemented solve users' problems.
- Users' acceptance of systems.
- Documentation of the systems.
- Ease of maintenance of the systems.
- Success in implementing systems on both time and cost schedules.
- Success in identifying likely areas for future development.
- Creation of good working relations with user departments.
- Technical efficiency of the systems (running times, core usage, etc.).

However, assessing the relative importance of these factors has so far proved impossible and to date there is still no formal technique available to management for judging the performance of systems analysts.

(5) There is a high degree of technological advance

One of the problems of systems management is to judge when it is best to take advantage of technological advances which are being made almost continually in the computer field. To try to keep up with every advance would incur the high costs of constant conversion. To stick at a given level of technical capability, on the other hand, gives a host of problems, such as maintenance of equipment, dissatisfaction of staff, decreasing cost/performance ratios in relation to what is available, etc. Even if the moment for change is judged correctly, the problems of conversion are great.

The above factors indicate some of the major problems which systems management face. Unfortunately, there is no simple solution, or set of solutions, to them. What is possible, however, is to apply known and successful management principles and practice to them. The major principles are:

- Use formal planning and control techniques (such as those described in Part IV, 21).
- Use the management science approach, especially cost-effectiveness techniques.
- Start now to assess and reward systems analyst performance.
- Do not try to be a computer pioneer.
- Apply the principles of modular systems (simplifies system development).

- Pay special attention to keeping staff both personally and professionally satisfied as they are the most valuable resource.

Only by following these and other principles can systems management retain control of the activities for which they are responsible. But retaining control is not of itself sufficient; one of the goals of systems management must be to contribute as much as possible to the company's success. So optimal use of resources is a requirement, as with the management of any other activity. This is particularly difficult in the multi-project environment, and for this reason systems management should make a special effort to use modern management techniques.

SELECTED BIBLIOGRAPHY

This selected bibliography is divided into two sections. Section 1 lists useful material to supplement the specific topics covered in this book. Section 2 presents more general material, which provides both basic reference works (such as a glossary) and texts indicating the wider implications of systems analysis and computer technology.

Section 1

Part One: The Six Steps of Systems Analysis

Laden, T. R., and Gildersleeve, H. N., *System Design for Computer Applications*, John Wiley & Sons Inc., New York, 1963.

Langefors, B., *Theoretical Analysis of Information Systems*, Student Literature, Lund, 1966.

Liston, D. M., and Schoene, Mary L., *A Systems Approach to the Design of Information Systems*, Journal of the American Society for Information Science, pages 115–22, March/April, 1971.

Part Two: Techniques

(1) *Fact gathering*

Millward, E., *Organisation and Methods*, Macmillan, 1959, second edition 1967.

(2) *Flowcharting*

Chapin, N., *Flowcharting with the ANSI Standard: A Tutorial*, Computing Surveys, Vol. 2, No. 2, June 1970, pages 119–46.

(3) *Decision Tables*

Fisher, D. L., *Data, Documentation and Decision Tables*, Communications of the ACM 9 (1966, 1), No. 1, pages 26–31.

King, P. J. H., *Decision Tables*, Computer Journal 10, 1967, pages 135–42.

King, P. J. H., *The Interpretation of Limited Entry Decision Table Format and Relationship Among Conditions*, Computer Journal, November 1969, pages 320–6.

——, *Decision Tables in Data Processing*, National Computing Centre, U.K.

(4) *Simulation*

Forrester, J. W., *Industrial Dynamics*, John Wiley & Sons Inc., New York, 1961.

Gordon, G., *System Simulation*, Prentice-Hall, 1969.

Koller, Dr. H., *Simulation als Methode in der Betriebswirtschaft*, Zeitschrift für Betriebswirtschaft, February 1966, No. 2.

Part Three: General Systems Considerations

Brandon, D. H., *Management Standards for Data Processing*, Van Nostrand, Princeton, N.J., 1963.

ed. Daniels, A., and Yeates, D., *Basic Training in Systems Analysis*, Pitman, 1969.

Dodd, G. G., *Elements of Data Management Systems*, Computing Surveys, Vol. 1, No. 2, June 1969, pages 117–33.

ed. Gentle, E. C., *Data Communications in Business—An Introduction*, Publishers' Service Company for American Telephone and Telegraph Company, 1965.

Kaufman, F., *Electronic Data Processing and Auditing*, Ronald Press, New York, 1961.

Martin, J., *Telecommunications and the Computer*, Prentice-Hall, 1969.

Martin, J., *Teleprocessing Network Organization*, Prentice-Hall, 1970.

——, *Richtlinien für die Revision Elektronischer Datenverarbeitung*, Wirtschaftsvereinigung Eisen- and Stahlindustrie Düsseldorf, Betriebswirtschaftlicher Ausschuss, 1960.

——, *Systems Documentation Manual*, National Computing Centre, U.K.

Mühring, W. J., *Controles bij Geautomatiseerde Informieverwerking*, Kantoor en Efficiency 5 (1966. 01), No. 44, pages 1838–42.

Part Four: Project Control

Lowe, C. W., *Critical Path Analysis by Bar Chart*, Business Books, 1966, second edition 1969.

Woodgate, J., *Planning by Network*, Business Publications, 1964.

Section 2

(1) *Glossaries*

Chandor, A., with Graham J., and Williamson, R., *A Dictionary of Computers*, Penguin Books, 1970.

——, *IFIP-ICC Vocabulary of Information Processing*, first English language edition, North Holland Publishing Company, Amsterdam, 1966.

(2) *Real Time Systems*

Latini, F., *Tempo Reale e Calcolatore Elettronico*, Etas Kompass, 1969.

Martin, J., *Design of Real Time Computer Systems*, Prentice-Hall, 1967.

(3) *Management Information Systems*

Blumenthal, S. C., *Management Information Systems: A Framework for Planning and Development*, Prentice-Hall, 1969.

Opitz, H., *Untersuchungen über die Einsatzmöglichkeiten von Datenverarbeitungsanlagen*, Westdeutscher Verlag, Köln, Opladen 1968.

Parsini, E., and Wächter, O., *Organisaţions—Handbuch für die Einführung von ADV-Systemen*, Walter de Gruyter Verlag, Berlin, 1971.

Rightman, R., Luskin, B. J., and Tilton, T., *Data Processing for Decision-Making*, Macmillan, second edition 1971.

Wilson, I. G., and Wilson, M. E., *Information, Computers and Systems Design*, John Wiley & Sons Inc., New York, 1965.

Yvon, P. J., and Semin, C., *Comment Concevoir un Système Intégré de Gestion*, Enterprise Modern d'Edition, 1971.

(4) *Implications of Automated Data Processing*

Diebold, J., *Man and the Computer*, Praeger, 1969.

Rose, M., *Computers, Managers and Society*, Penguin Books, 1969.

Warner, M., and Stone, M., *The Data Bank Society*, George Allen and Unwin, 1970.

(5) *General Reference*

——, *Information Systems Handbook*, Philips Data Systems, 1968.

INDEX

Absolute checks 125
Accounts receivable system 88
ALGOL 132
American National Standards Institute (ANSI) 132
Amplitude modulation (AM) 154–5
ANSI-COBOL 129
Application packages 174
Arrow diagrams 177
Asynchronous transmission 154
Audit controls 127
Automated flowcharting 58–60

Bad debt values 16
Bar charts 176
Baud 157
Benchmarks 72–4
Benefits 15
Binary search 100
Blocking 118
Brainstorming 30
Break-even points 20
British Standards Institution (BSI) 132
Bursting 93

Capital assets system 17
Card codes 82
Card readers 88
Card records 97–8
CASE 74
Cassette tapes 91
Character checks 126
Character printers 93–4
Characters 88
Check digits 124
Checkpoints 37–8
COBOL 65, 72, 147–8
Codes 85, 89–90, 160
Coding systems 160
Common data base 94, 115, 118, 143
Communications networks 75–7, 152, 159–60
Compiler 65
Computers 1, 12, 88, 90–1, 93
Computer manufacturers 133

Computer programs 12, 27, 90
Computer room procedures 120
Control totals 123
Costs 12–14, 35
Cost accounting system 6
Cost-benefit analysis 11–12, 20, 35
Criteria for project selection 8
Critical path method (CPM) 177
Cylinder concept 110

Data bank 94, 115, 118, 143
  bases 94, 115, 118, 143
  capture 81–93, 153–5
  collection and transmitting equipment 153–5
  communications 74–5, 152, 158–60
  compaction 112–14
  control 85–6, 123
  element description 143–9
  elements 24, 28, 115
  preparation 85
  security 81, 120–7
  sequence checks 125
  structures 115
  transfer rate 88
Debugging programs 33–4, 59–60
Decision tables 27, 61–7
Decision tables, comparison with flowcharts 67
Decision table form 140
De-collating 93
Design considerations 81–4, 158–60
  ideas 29–30
  phase 5, 13, 26
  specification 31
Desk simulation 71
Dictionary 101–2
Digital-analogue converter 95
Direct access 98, 114, 119
Discounted cash flow 19
Documentation 45, 56, 58, 137–51
Document counts 123
Dumping files 114
Duplex (full duplex) 156

189

Editing   113
Encoding   112
Enforcement of standards   133–6
Envelope   154
Error correction   90
Error rates   35, 81, 123–6, 156
European Computer Manufacturers' Association (ECMA)   132
Evaluation phase   5, 13, 35–6
Exception data   82
Exception reporting   83–4
Existing system   37

Fact gathering   44–51
Feasibility study   5, 8, 11, 13
Field   115
File(s)   87, 89, 109, 117–18
   conversion   174–5
   creation   171, 174–5
   interrogation   28–9, 31, 111
   organisation   99–104, 172
   security   114
   specification sheet   143
Flowcharting   52–60, 66–7
   form   140
   symbols   54–5
Forms   139–51
FORTRAN   75
Frequency modulation (FM)   154–5

GANTT charts   176–80
GASP (Graphical procedure for Analytical and Synthetical evaluation and review of construction Programs)   179–81
Generalised file interrogation   28–9, 31, 111
Glossaries   149–50
GPSS   74–5
Grandfather, father, son principle   114
Group   115

Half duplex   156
Hardware   6, 12, 107
Hardware controls   121
Hash totals   123
Header labels   126
Human relations element   43

Implementation phase   5, 13, 33
Indexed sequential system   100
Informal standards   133
Input/output   81–95
Insurance system   17
Intangible benefits   15
Integrated systems   7, 9
International Standards Organisation (ISO)   52, 132, 140

Interviewing   44–6, 49–51, 64
Inverted list structure   104
Invoicing system   47
Item   115

Key   99, 101–3, 118–19
Keypunching   82, 85
Keytaping   90–1
Key-to-disk   91–2
Kimball tags   90

Library control   120
Limitation of personnel   120–1
Line printers   93
List structures   103
Logical organisation   119
Logical record   116

Magnetic disks   12
Magnetic ink character recognition (MICR)   92
Magnetic tape   12, 90–1, 95
Maintenance   161
Management   135, 167, 182–5
   involvement   8, 166–8, 182–5
Manual of standards   134
Mark sensing   86–8
Message volume   159–60
Microfilm   94
Mini-computers   93
Modification procedure   163
Modularity   27–9, 162
Modular systems   27–9
Modulus checks   124
MPM (Metra Potential Method)   177

National Computing Centre (NCC)   132
Natural hazard protection   121
Network charts   176, 179–80
Net present value (NPV)   19
Noise level   90

Observation   47
Off-line printers   93–5
Operating systems   109, 172
Optical character reading (OCR)   69, 91
Order entry systems   7, 17, 23
Organisation   164
Organisation and methods   1
Output   11, 81–3, 93
Overflow areas   102, 114

Packing   113
Paper tape   83, 89, 94
Parallel running   171, 174
Parameters   28–31
Pareto's Law   73

INDEX

Parity 89, 121, 154
Partially inverted list structure 104
Payback period 18
Payroll system 82
Peak loads 34, 73
Performance standards 131
Personnel records system 57
Personnel statistics system 58–9
PERT (Program Evaluation and Review Technique) 176
Phase modulation (PM) 154–5
Physical organisation 100, 119
Physical security 120–1
Picture checks 126
Planning and control 33, 175–81
Plotters 94
Pointers 103
Points weighting schemes 17, 71–2
Polling 158
Precedence charts 177
Preprocessors 64
Printer layout chart 141
Procedures 64, 82, 120
Procedure description sheet 139
Production information 61
Production planning system 24
Production scheduling system 7
Programming 109, 175
Program specifications 27
Project selection 5–6
Punched cards 82, 85–9, 94
Punched card layout form 141
Purging files 113

Quality control system 9
Questionnaires 48

Random organisation 101–2
Reasonableness checks 124
Receiving units 157
Recording facts and data 46, 48–9, 85
Records 89, 98, 116
Record layouts 82
Record size 110
Response times 11, 111, 160
Return on investment 8, 18
Ring structure 104

Sales accounting systems 28
Sampling 47
Sampling checks 126
Savings 15
SCERT 74
Selection of computer system 71–7

Sequence checks 125
Sequential access 119
Sequential organisation 99–100, 115
Simplex 156
SIMSCRIPT 74–5
Simulation 68–77
Slave computers 93
Software 6, 27, 60, 65, 74
Sorting 85, 88
Sources of standards 132
Space management 112
Spooling 83, 93
Standards, definition of 128
Standards manual 134
Standard report form 139–40
Statistics 47
Steering committee 182
Stock control system 68, 70, 83
Storage devices 119
Storage media 119
Sub-systems 23
Supervisory checks 123–4
Synchronous transmission 154
System definition 5, 13, 22, 24–5
    design 26–32
    maintenance 161
    objectives 2, 35–6
Systems controls 121–7
Systems modification request 151, 163

Tape records 98
Telex 89
Terminals 92, 153–4
Track 110
Trailer labels 126
Training 173–4
Training, systems analyst 43
Training, users 34, 173–4
Transport and collection of data 85
Trimming 93
Turnaround document 88
Turnaround times 91

Updating 111
User groups 132–3
User involvement 162, 166–8

Value of information 16
Variable length records 113, 116–7
Verification 85, 88–90, 92, 124–5
Visual display devices 8, 94

Weighting and ranking 72
Working documents 88